The Rayleigh–Ritz Method for Structural Analysis

To my late parents Saraswathyppillai and Sinniah, my late brother Senthinathan who encouraged and supported me during my studies at Manchester, my brothers Kumarabharathy, Kathirgamanathan, sister Sooriyakumari, my wife Krshnanandi, daughters Kavitha and Tehnuka, my in-laws, nephews, nieces, my supervisors Emeritus Professor Dickinson, the late Dr Tillman, my teachers from my old schools in Sri Lanka (Veemankamam Mahavithiyalaym, Mahajana College, Tellippalai), and all my lecturers and students, and colleagues both current and past.

Sinniah Ilanko

To my wife and son.

Luis Monterrubio

Series Editor
Noël Challamel

The Rayleigh–Ritz Method for Structural Analysis

Sinniah Ilanko
Luis E. Monterrubio

with editorial assistance from
Yusuke Mochida

iSTE

WILEY

First published 2014 in Great Britain and the United States by ISTE Ltd and John Wiley & Sons, Inc.

ISTE Ltd
27-37 St George's Road
London SW19 4EU
UK

www.iste.co.uk

John Wiley & Sons, Inc.
111 River Street
Hoboken, NJ 07030
USA

www.wiley.com

Library of Congress Control Number: 2014953191

British Library Cataloguing-in-Publication Data
A CIP record for this book is available from the British Library
ISBN 978-1-84821-638-9

Contents

Preface

It is a privilege to have the opportunity to share with you some of our interesting experience in the journey of learning and research about structural analysis. The specific path we are exploring here is a superhighway – a variational technique called the Rayleigh–Ritz method or the Ritz method. I was introduced to this method during my PhD studies by an excellent supervisor, Professor Stuart Dickinson at the University of Western Ontario. Prior to this, another excellent supervisor (my BSc and MSc supervisor), the late Dr Stuart Tillman (University of Manchester), had taught me the Lagrangian multiplier method. Dr Tillman's lectures were such that one has only to listen once – from then on the material stays crystal clear. So the reason I became interested in these techniques (other than the fact that they are very handy for vibration analysis which was my research area) is perhaps the passion that my teachers showed in the subject.

The thought of writing a book in vibration occurred to me after my first study leave in India where I received some positive feedback after giving a public lecture on "Vibration and Stability of Structures" that was meant to be for a general audience. I tried hard to think of ways of explaining concepts such as natural frequency, stiffness, mass, damping where possible using everyday experiences and analogies. While such analogies may not be accurate, they help to create an image and this technique has since helped me to score some points with my students. For example, I explain natural frequencies as the frequencies at which a structure is easily excitable and give as an example the frequency at which one should meet with a potential partner or friend to sustain the relationship. If we meet too frequently, we may not give our friends enough space and scare them; if do not meet often enough, we may wrongly signal that we are not interested in them. So to get the maximum response, we need to engage at the right frequency. The same goes with structures. I have always wanted to share such thoughts through a book.

The opportunity to write a book has finally come through an invitation from Professor Noël Challamel. The book is about the Rayleigh–Ritz method but as you will see, for historical reasons and for its common potential use, the focus is largely on natural frequencies and modes and the related problem of structural stability. I have tried to think of simple analogies to present this in an interesting way but have only managed to do this in one or two places. The book is a mixture of well-established material (both theory and application) and the result of our own research. An accidental mistake in the sign of an inertia term in an equation in my PhD thesis has led to the discovery that negative values of large magnitude for masses and stiffness can be used in modeling constraints. Dr Luis Monterrubio has utilized this idea for solving a number of different structural problems, and has come up with a nice set of admissible functions to use in the analysis of beams, plates, shells and solids in Cartesian coordinates and has contributed to Chapters 8–12. Dr Yusuke Mochida has helped with proofreading and checking the derivations. I am grateful to Luis, Yusuke, Professor Noël Challamel and ISTE for making this possible.

I must also record my thanks to my family, teachers, students and colleagues for my role as an author would not have been possible without them.

Some of our explanations, for example the use of analogies, may not be based on principles of science, or may be in subject areas such as management in which we do not have any expertize. While we have tried to eliminate errors, our attempts to continuously improve the manuscript with new examples and explanations may have led to some mistakes. We would be grateful to receive any comments, criticisms or suggestions for improvements regarding the contents of this book.

Sinniah ILANKO

October 2014

Introduction and Historical Notes

In many practical engineering problems, it is not possible or convenient to develop exact solutions. A convenient method for solving such problems originated from attempts to calculate natural frequencies and modes of structures. This method is known as the Rayleigh–Ritz method or the Ritz method [RAY 45a, RAY 45b, RIT 08, RIT 09, LEI 05, ILA 09, YOU 50]. In this book we will see how to apply this method for solving a variety of common problems engineers and scientists encounter. We will first provide some historical notes on the development of the method and show how the principle of conservation of energy leads to this procedure. Those who are keen to get on with the application may want to proceed to Chapter 3.

Chapter 1 starts with application of the principle of conservation of energy for a simple pendulum showing how the natural frequency can be found by applying this principle for a system that can vibrate only in one mode or shape. Such systems that can only vibrate in one mode are called single degree of freedom systems, as a single coordinate is sufficient to describe the actual shape of natural vibration and such a natural vibration without any external dynamic force takes place only at one frequency which is its natural frequency. Then we consider a spring-mass vibratory system which has two independent coordinates. It can be easily shown, by applying Newton's second law of motion, that the system has two natural frequencies with associated modes. Application of the conservation of energy for this system requires an assumption about the shape of vibration. Although the natural frequencies and modes can be calculated conveniently by solving the equations of motion derived from Newton's second law, application of the principle of conservation energy shows that in this case this the application leads to one value for the frequency which depends on the assumed mode (the assumed ratio of the displacement of two masses) and takes minimum and maximum values when the assumed modes correspond to the actual first and second modes, yielding the respective natural frequencies. In Chapter 2, we proceed to show how this illustrates Rayleigh's Principle which in Lord Rayleigh's own words is stated as follows:

> The period of a conservative system vibrating in a constrained type about a position of stable equilibrium is stationary in value when the type is normal.

Chapter 2 also presents a well-known proof that for a system with a finite number of degrees of freedom, the frequency obtained by applying the principle of conservation of energy is an upper bound to the fundamental natural frequency, and a lower bound to the highest natural frequency, provided no essential (geometric) constraints are violated. The requirement that the essential conditions are not violated leads to the notion of admissible forms (these could be vector or functions) of displacements. This chapter also deals with what is meant by admissibility.

Lord Rayleigh has shown in several of his papers and books, how a good estimate for the fundamental natural frequency may be obtained by adjusting the shape of a chosen function to seek the lowest possible values for the frequency (or highest value for the period). The expression for the frequency is a quotient with potential energy being the numerator and a kinetic energy function being the denominator. This quotient is called the Rayleigh quotient. However, the credit for introducing a systematic method for performing this minimization should be given to Walter Ritz. For this reason there are some who argue that the method should be called the Ritz method. The arguments and counterarguments for the name are available in literature [LEI 05, ILA 09]. So we will not focus on it here except to say that to be inclusive we are using the name Rayleigh–Ritz method which gives credit to both Rayleigh and Ritz.

Thus, Chapter 3 takes us from the implication of Rayleigh's principle to the Rayleigh–Ritz method. Typical minimization equations are formulated for a conservative structural system possessing potential and kinetic energy, to the point of developing the eigenvalue equations. A cantilever beam is used as an illustrative example showing how the method is applied to obtain the natural frequencies and modes. The effect of adding partial restraints and rigid body attachments are also explained in this chapter. In addition to natural frequency calculations, static analysis is also demonstrated as a special case. It may be worth noting here that while the origin of the Rayleigh–Ritz method can be traced back to problems of finding natural frequencies and modes, it can also be used to solve boundary value problems. In structural analysis, this corresponds to calculation of displacements using the minimum total potential energy theorem, but the procedure for minimization is the same as the one used in the Rayleigh–Ritz method for vibration analysis.

We have noted that a requirement of the Rayleigh–Ritz method is that the choice of displacement functions for formulating the energy terms is subject to the requirement that they satisfy all geometric conditions. In actual fact, it is not necessary for each function to satisfy the constraints but the series as a whole does need to. A way to relax this requirement is to use the Lagrangian Multiplier method where each function is allowed to violate the geometric constraints but then these constraints are enforced by including additional constraint equations which are associated with undetermined coefficients called the Lagrangian Multipliers. Chapter 4 deals with this approach and demonstrates the method through a propped cantilever.

Chapter 5 presents some mathematical derivations and formulas for computing the terms in the eigenvalue matrix equations. For example, integral expressions for stiffness and mass matrices are presented in Chapter 5.

Chapter 6 introduces the penalty method. While the Lagrangian Multiplier method helps to relax the limitations on the choice of admissible displacement functions or vectors, it introduces extra equations that need to be solved together with a set of minimization equations. There is another clever way to achieve the enforcement of geometric conditions without increasing the number of equations. This involves a gemoetric constraint with an artificial spring of very high stiffness and including the strain energy associated with any violation of the constraint. This idea was introduced by Richard Courant in [COU 43] and has since then become very popular and widely accepted. This is known as the penalty method. The penalty parameter corresponds to the stiffness of the artificial spring and serves as a penalty against any constraint violation. There have been two criticisms about this approach. One is that while high stiffness may minimize any constraint violation, it is not possible to determine the effect it has on the accuracy of the results. Furthermore, choosing a stiffness that is large enough to prevent any constraint violation, but not too large as to cause any numerical problems due to round-off errors, can be challenging. In the case of frequency calculations, the approximation of a rigid boundary condition with a less than ideally rigid condition relaxes the structure and could result in lower estimates for the natural frequencies. This means that the Rayleigh–Ritz method would then yield an upper bound solution to a lower bound model. However, recent advances in the penalty method where stiffness parameters of positive and negative values had been used were found to give bounded results for frequencies, as far as the constraint violation is concerned. This is explained in Chapter 6.

Although it is well known that the Rayleigh–Ritz method gives upper bound to the fundamental natural frequencies, it is not well known that the method actually gives upper bound to all but the highest of the natural frequencies. The proof of boundedness of the Rayleigh–Ritz method, which comes in the form of Theorem of Separation, is presented in Chapter 7. This chapter also gives rigorous mathematical proof of theorems that justify the use of negative stiffness parameters, which were derived by one of the authors.

The fact that the admissible functions do not have to satisfy all geometric constraints is great news for the Rayleigh–Ritz method fans because this removes the restrictions on the choice of admissible functions. However, with the use of penalty terms, some functions are known to cause numerical problems. So, are there any well-behaved functions? This question is answered in Chapter 8, where a recipe for formulating shape functions can be found, presenting a specific set for many common structural elements including beams, rectangular plates, shells of rectangular planform and solids – all are shapes which can be described conveniently in Cartesian coordinates. It is a convenient set consisting of a cosine series, and linear and quadratic functions. These functions are easy to work with and have shown to be well-behaved even under challenging conditions such as high penalty terms and higher modes where most common admissible functions cause numerical problems.

Chapters 9 deals with application of the special set of functions for beams, which is then extended to rectangular plates in Chapter 10, shells of rectangular planform in Chapter 11 and solid bodies in Chapter 12. In all these cases, it has been shown that the set of admissible functions presented in Chapter 7 can be used to find the natural frequencies and modes of the totally unconstrained system, without causing any numerical problems. This is observed consistently for beams, plates, shells and solids, irrespective of the number of terms used or the number of natural frequencies and modes obtained. For any other set of boundary conditions, the penalty method has been employed to obtain the natural frequencies and modes.

Chapter 13 deals with an interesting topic in vibration, namely the effect of a static axial force on the natural frequencies of a beam. Although the focus of the book is on the Rayleigh–Ritz method, we have taken the opportunity to discuss the relationship between the natural frequencies and critical loads of a structure. Anyone who has played any musical string instrument will know that increasing the tension increases the pitch. This is because the stiffness increases with tension. Those who have not had the luxury of handling a string instrument may be able to attest to this by doing a simple experiment with a clothesline. If the tension is increased it is harder to move the clothesline and it would sag less under the weight of the clothes. This is because the tension increases the stiffness which would give rise to an increase in potential energy and natural frequency. With compressive force the stiffness decreases and the frequencies also decrease. The interesting point is that if a natural frequency approaches zero then its period would tend to infinity, meaning that if the structure is displaced slightly and released it would take infinite time to return to its equilibrium state, meaning it will never return to equilibrium. This point of no return is the critical state and it is signalled by a frequency approaching zero.

Having dealt with analytical methods, we conclude the book with Chapter 14 which explains how the Rayleigh–Ritz method is used as the basis of the ubiquitous Finite Element method. It is simple: change the name of "admissible functions" to "shape functions" and derive the system matrices for a structural element and then assemble these matrices by using common values of the displacement and or slope at the nodes where they are connected. This gives us the Finite Element method. We admit this is an over simplification, but this chapter shows how the FEA matrices can be developed for beam elements as we felt it would be nice way to end the book by showing its connection to another subject.

1

Principle of Conservation of Energy and Rayleigh's Principle

The well-known principle of conservation of energy forms the basis of some common convenient analytical techniques in Mechanics. According to this principle, the total energy of a closed system remains unchanged. This means that in the absence of any losses due to friction etc., the sum of the total potential energy and the kinetic energy of a vibratory system will be a constant. Although in practice there will always be some damping, and hence dissipation of energy, for many mechanical systems such losses may be neglected. Such systems are called conservative systems.

The natural frequencies of conservative systems may be obtained by equating the maximum kinetic energy (T_m) to the maximum total potential energy (V_m) associated with vibration. The meaning of these energy terms is very important. To illustrate the principle of conservation of energy, and the meaning of the energy terms let us study some simple vibratory systems.

1.1. A simple pendulum

Consider the oscillatory motion of the simple pendulum consisting of a bob of mass m and a massless string of length L as shown in Figure 1.1. It would be at rest in a vertical configuration under gravity field. If it is given a small disturbance β_m and then released, it will tend to vibrate about this equilibrium state. The restoring action of the gravity force will initiate a motion toward the equilibrium state but as the bob approaches the lowest point in its motion it has a velocity and therefore carries on swinging up on the other side until the gravity force causes it to come to a halt momentarily. In the absence of any damping forces, this motion would go on forever, but in reality the damping forces will help to put an end to this vibration after some time.

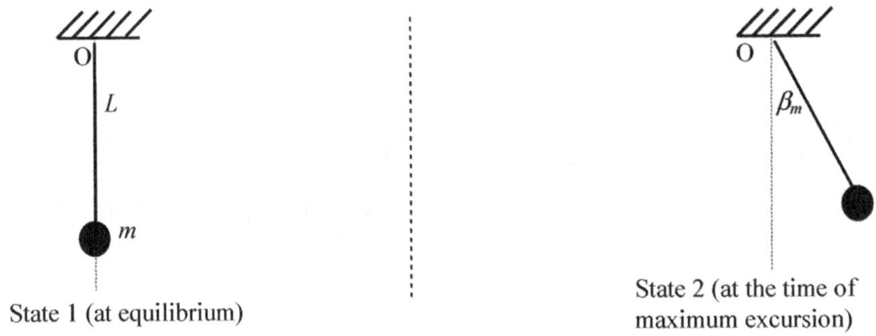

Figure 1.1. *Simple pendulum*

Assuming that energy loss associated with mechanical friction and aerodynamic resistance is negligible, we have two types of energy term to consider. These are the kinetic energy (denoted by T_1, T_2) where the subscripts 1 and 2 refer to states 1 and 2 respectively, and the potential energy (denoted by V_1, V_2). The kinetic energy is proportional to the square of the velocity and the potential energy is dependent on the vertical position of the bob.

The pendulum will have the maximum kinetic energy as the bob passes through the equilibrium state (state 1) at which time it will have the lowest potential energy. At the time of maximum excursion (state 2), the bob will be at its highest point, and therefore the system will have the maximum potential energy, but since it has no velocity its kinetic energy will be minimum. The potential energy can be defined arbitrarily by selecting a datum. In our example, the increase in the potential energy as the system changes from state 1 to state 2 is entirely associated with the vibration, and will be referred to as the maximum potential energy hereafter. As the bob returns to state 1 from state 2, it loses potential energy and gains kinetic energy. The maximum kinetic energy associated with vibration is the kinetic energy at state 1 minus the kinetic energy at state 2. (The latter is not necessarily absolutely zero, as the support point may have a velocity. In rotating systems care must be taken to ensure that the kinetic energy terms are calculated correctly.) Since the total energy is conserved, the maximum kinetic energy associated with vibration must be equal to the maximum total potential energy associated with vibration. The inclusion of the phrase "associated with vibration" is used here since terms such as "maximum" and "total" can otherwise cause confusion.

From the principle of conservation of energy:

$$V_1 + T_1 = V_2 + T_2$$

i.e. $V_2 - V_1 = T_1 - T_2$

The gain in potential energy as the bob moves from state 1 to state 2 is the maximum potential energy associated with vibration and may be denoted by V_m.

$$V_2 - V_1 = V_m$$

Similarly the maximum kinetic energy associated with vibration is:

$$T_1 - T_2 = T_m \quad [1.1]$$

From the above equations we have $V_m = T_m$

In applying the principle of conservation of energy for vibratory systems, it is sufficient to equate the maximum potential and kinetic energy terms associated with vibration.

To find the circular natural frequency ω of an undamped system, the motion may be assumed to be simple harmonic.

i.e. $\beta = \beta_m \sin(\omega t + \alpha)$, where t is time and α is a phase shift angle.

Then $\dfrac{d\beta}{dt} = \dot{\beta} = \omega\beta_m \cos(\omega t + \alpha)$

The maximum velocity is therefore $= L\omega\beta_m$

This means the maximum velocity is equal to the amplitude of vibration times the frequency. This statement is true for any natural mode, since at natural modes the vibration is simple harmonic.

Hence $T_m = m\dfrac{(L\omega\beta_m)^2}{2}$

The potential energy is due to the change in position of the gravity force mg.

Thus $V_m = mgL(1 - \cos\beta_m)$

Substituting these into equation [1.1] gives:

$$m\frac{(L\omega\beta_m)^2}{2} = mgL(1-\cos\beta_m)$$

For small amplitude vibration, $(1-\cos\beta_m) = \frac{\beta_m^2}{2}$

This gives: $m\frac{(L\omega\beta_m)^2}{2} = mgL\frac{\beta_m^2}{2}$

This actually gives us two possible solutions. One is that $\beta_m = 0$. This implies there will not be any motion and is therefore a trivial solution. The other solution is:

$$\omega^2 = g/L$$

in which case $\beta_m \neq 0$ and vibration is possible. That is to say, in the absence of any external force, the system can vibrate freely at a frequency of $\omega = \sqrt{g/L}$ rad/s. This is therefore the circular natural frequency of the pendulum. The natural frequency in Hz (cycles/s) is $f = \omega/2\pi = \sqrt{g/L}/2\pi$. From this point onwards, for simplicity, we will refer to circular natural frequencies as natural frequencies.

The simple pendulum is a "single degree of freedom" system. This means it can only vibrate in one specific mode, in this case, the string and the bob rotating about the equilibrium state. In cases where the mode is defined, the above method yields the exact value of the natural frequency. We will soon see why it is not always possible or convenient to get the exact frequency.

1.2. A spring-mass system

Consider the motion of a simple spring-mass system shown in Figure 1.2. A rigid body of mass m is connected to a linear elastic spring of stiffness k. Assume that the system is free to vibrate only axially (in the direction of the spring). If the mass is displaced from the equilibrium state by distance $\hat{u}$ which induces a force in the spring and then released, it would tend to return to its equilibrium state. However as it approaches the equilibrium state it has a velocity and this velocity causes the mass to move away from the equilibrium state now on to the opposite side. Then as the spring force develops, the motion comes to an end momentarily and the mass then returns to the equilibrium state and the cycle repeats as in the case of the pendulum.

The maximum vibratory potential energy in the spring is given by $V_m = \dfrac{k\hat{u}^2}{2}$ and the maximum kinetic energy is $T_m = \dfrac{m\omega^2\hat{u}^2}{2}$.

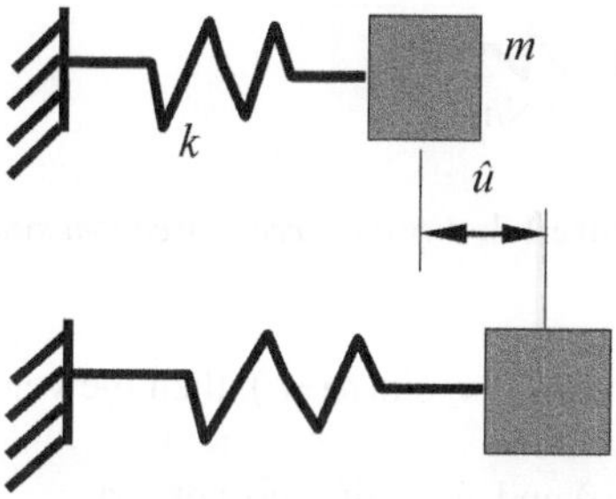

Figure 1.2. *A spring-mass system*

Using equation [1.1], $\dfrac{k\hat{u}^2}{2} = \dfrac{m\omega^2\hat{u}^2}{2}$

For a non-trivial solution, we have the natural frequency given by $\omega = \sqrt{k/m}$.

For multidegree of freedom systems and continuous systems, calculation of the energy terms requires an assumption about their modes. This method results in exact values of natural frequencies only if the exact modes are used in the calculations. If the exact mode is not known, it has been shown that the use of any other mode shape that does not violate any geometric constraints of the system results in a frequency that cannot be lower than the exact fundamental natural frequency. This is one of the many interesting proofs that Lord Rayleigh gives in his famous book *The Theory of Sound* [RAY 45a]. This is best illustrated through some examples. We will start with a two degree of freedom system, as its exact solution is easily understood.

1.3. A two degree of freedom system

Consider the vibration of the spring-mass system shown in Figure 1.3. The masses are assumed to be free to move only in the axial direction. Since the two masses can be moved independently of each other there are two degrees of freedom. The relevant values for the stiffness and mass are shown in the figure. Let the dynamic displacement of the masses be u_1, u_2 and their amplitudes be $\hat{u}_1$, $\hat{u}_2$. The various energy terms may be written in terms of $\hat{u}_1$ and $\hat{u}_2$.

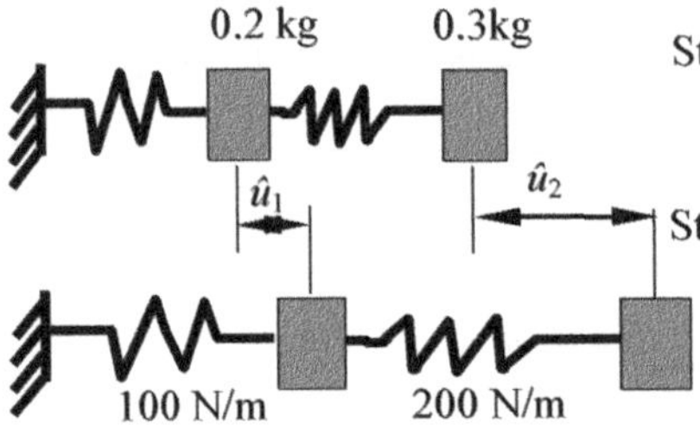

Figure 1.3. *A two degree of freedom system*

If $u_1 = \hat{u}_1 \sin(\omega t+\alpha)$ and $u_2 = \hat{u}_2 \sin(\omega t+\alpha)$, then the velocities are given by:

$$\dot{u}_1 = \omega \hat{u}_1 \cos(\omega t+\alpha) \text{ and } \dot{u}_2 = \omega \hat{u}_2 \cos(\omega t+\alpha).$$

The maximum potential energy due to vibration is

$$V_m = (1/2)100\,(\hat{u}_1)^2 + (1/2)200\,(\hat{u}_2 - \hat{u}_1)^2$$

The maximum kinetic energy is:

$$T_m = (1/2)(0.2)\,\omega^2 \hat{u}_1^{\,2} + (1/2)(0.3)\,\omega^2 \hat{u}_2^{\,2}$$

Using the principle of conservation of energy, $V_m = T_m$.

This gives:

$$(1/2)100\,(\hat{u}_1)^2 + (1/2)200\,(\hat{u}_2 - \hat{u}_1)^2 = (1/2)(0.2)\,\omega^2 \hat{u}_1^{\,2} + (1/2)(0.3)\,\omega^2 \hat{u}_2^{\,2}$$

$$\text{i.e. } \omega^2 = [(1/2)100\,\hat{u}_1^{\,2} + (1/2)200\,(\hat{u}_2 - \hat{u}_1)^2]/[\,(1/2)(0.2)\,\hat{u}_1^{\,2} + (1/2)(0.3)\,\hat{u}_2^{\,2}]$$

This equation, which is the Rayleigh Quotient, can be condensed into the following form:

$$\omega^2 = (100\,\eta^2 + 200\,(1-\eta)^2)/(0.3 + 0.2\,\eta^2), \qquad [1.2]$$

where $\eta = \hat{u}_1/\,\hat{u}_2$ (the mode). This is equivalent to normalizing the amplitude of the displacements of the masses with respect to $\hat{u}_2$. Note that modes give a relationship between the displacements of the degrees of freedom in the system.

Unlike the first example, this does not immediately yield a value for the natural frequency. However, it may be solved by assuming a relationship between $\hat{u}_1$

and $\hat{u}_2$ (i.e. assuming a mode). The exact natural frequencies and modes found by solving the two equations of motion (obtainable by applying Newton's second law) are:

$$\omega_1 = 12.97 \text{ rad/sec}, \ \hat{u}_1/\hat{u}_2 = 0.75 \text{ and } \omega_2 = 44.72 \text{ rad/sec}, \ \hat{u}_1/\hat{u}_2 = -2$$

Substituting the fundamental mode into the frequency equation gives the exact result for the fundamental natural frequency. *Any other value for η results in a higher value for the frequency* (see Figure 1.4). This is an important observation and will be discussed later.

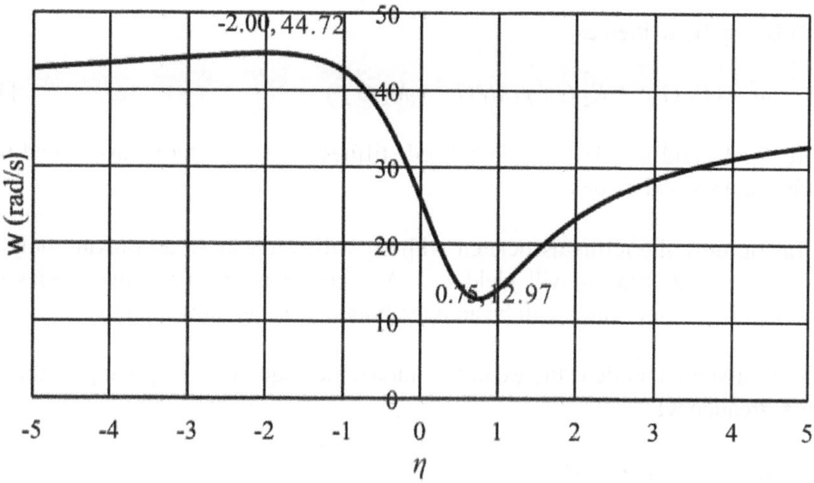

Figure 1.4. *Variation of calculated frequency with trial mode*

It is possible to show that the frequency calculated will not be lower than the fundamental natural frequency. We will do this by expressing the assumed mode in terms of the natural modes.

$$\begin{Bmatrix} \hat{u}_1 \\ \hat{u}_2 \end{Bmatrix} = G_1 \begin{Bmatrix} 0.75 \\ 1.00 \end{Bmatrix} + G_2 \begin{Bmatrix} -2.0 \\ 1.00 \end{Bmatrix}$$

Then the total potential energy is:

$$V_m = (1/2)(100)(0.75\, G_1 - 2\, G_2)^2 + (1/2)(200)(\,(0.75\, G_1 - 2\, G_2) - (G_1 + G_2))^2$$

$$= 34.375\, G_1^{2+}\, 1100\, G_2^{\,2}$$

$$T_m = (\omega^2/2)(0.2)(0.75\,G_1 - 2.0\,G_2)^2 + (\omega^2/2)(0.3)(G_1 + G_2)^2$$

$$= \omega^2(0.20625\,G_1^{\,2} + 0.55G_2^{\,2})$$

It may be noted that the energy terms do not contain the product G_1G_2 and this is due to the orthogonality of the natural modes:

$V_m = T_m$ gives

$$\omega^2 = \frac{34.375G_1^2 + 1100G_2^2}{0.20625\,G_1^2 + 0.55G_2^2} \qquad [1.3]$$

This may be written as

$$\omega^2 = (K_1\,G_1^{\,2} + K_2\,G_2^{\,2})/(M_1\,G_1^{\,2} + M_2\,G_2^{\,2}) \qquad [1.4]$$

where K_1, K_2 and M_1, M_2 are generalized stiffness and mass terms associated with the first and second modes.

Taking only the terms associated with G_1 will result in ω_1 and taking only the terms associated with G_2 will yield ω_2. Any combination of the two modes will result in a frequency which will be an upperbound to ω_1 as shown below.

If the assumed mode is the exact first mode, then equation [1.4] will give the first natural frequency;

i.e. $\omega_1^{\,2} = K_1\,G_1^{\,2}/\,M_1G_1^{\,2} = K_1/\,M_1$

Therefore,

$$K_1 = M_1\omega_1^{\,2} \qquad [1.5]$$

Similarly it can be shown that

$$K_2 = M_2\omega_2^{\,2} \qquad [1.6]$$

Substituting equations [1.5] and [1.6] into equation [1.4] gives:

$$\omega^2 = (M_1\,\omega_1^{\,2}\,G_1^{\,2} + M_2\,\omega_2^{\,2}\,G_2^{\,2})/(\,M_1\,G_1^{\,2} + M_2\,G_2^{\,2})$$

Dividing the numerator and denominator of the RHS of this equation by $\omega_1^{\,2}\,G_1^{\,2}$ gives:

$$(\omega/\omega_1)^2 = (M_1 + M_2(\omega_2/\omega_1)^2\,(G_2/G_1)^2)/(\,M_1 + M_2(G_2/\,G_1)^2)$$

Since $\omega_2 \geq \omega_1$, on the R.H.S., the numerator $\geq$ denominator. Hence $\omega \geq \omega_1$.

This means that the calculated frequency will be an upperbound to the fundamental natural frequency. In the next chapter we will extend this proof for a system with any number of degrees of freedom, and discuss Rayleigh's principle. We will then see that this statement of boundedness is true for all conservative systems, as long as the assumed displacement configuration satisfies certain conditions as explained in the next chapter.

2

Rayleigh's Principle and Its Implications

2.1. Rayleigh's principle

In Article 88 of *The Theory of Sound* [RAY 45a], Rayleigh gives the following theorem:

> The period of a conservative system vibrating in a constrained type about a position of stable equilibrium is stationary in value when the type is normal.

The meaning of "normal" type is that it is a natural mode. This statement, known as Rayleigh's principle has been given the following interpretation by Temple and Bickley [TEM 33]:

> In the fundamental mode of vibration of an elastic system, the distribution of kinetic and potential energies is such as to make the frequency a minimum.

In the context of its use in the Rayleigh–Ritz procedure, we could use the following interpretation:

> Equating the maximum total potential energy associated with vibration to the maximum kinetic energy associated with vibration results in an upperbound estimate of the fundamental natural frequency, provided the dynamic displacement forms assumed are admissible (permissible). A displacement function is admissible if it does not violate any geometric constraints and can represent the

displaced form of the system without any discontinuity. (If a system can be given an initial displacement in a given form that form is admissible).

The proof is essentially as mentioned in the next section.

2.2. Proof

Since the admissible functions do not violate any of the geometric constraints of a given system, it is possible to express any choice of admissible function as a linear combination of the system's natural modes. Let f be the assumed displacement form at the time of maximum excursion (which may be a vector for a discrete system or a continuous function for a continuous system). Let ϕ_i be the ith mode of the system (again for discrete system this would be a set of co-ordinates or a vector, and for continuous systems a function).

$$f = \sum_i G_i \phi_i \qquad [2.1]$$

Since the natural modes are orthogonal, the potential and kinetic energy terms may be written in the form:

$$V_m = \sum_i K_i G_i^2 , \qquad [2.2]$$

and

$$T_m = \omega^2 \sum_i M_i G_i^2 , \qquad [2.3]$$

where K_i, M_i are the generalized stiffness and mass terms respectively.

Equating the two energy terms results in the following equation for the frequency:

$$\omega^2 = \frac{\sum_i K_i (G_i)^2}{\sum_i M_i (G_i)^2} \qquad [2.4]$$

If the assumed displacement f is the ith natural mode, then equation [2.4] would give the ith natural frequency.

i.e. $\omega_i^2 = \dfrac{K_i (G_i)^2}{M_i (G_i)^2} = \dfrac{K_i}{M_i}$

giving

$$K_i = M_i \omega_i^2 \quad [2.5]$$

Substituting [2.5] into equation [2.4] gives:

$$\omega^2 = \frac{\sum_i M_i \omega_i^2 (G_i)^2}{\sum_i M_i (G_i)^2} \quad [2.6]$$

Dividing the numerator of both sides by ω_1^2 and dividing the numerator and the denominator of the R.H.S. by $M_1 (G_1)^2$ gives the following expression which is referred to as the Rayleigh quotient:

$$\left(\frac{\omega}{\omega_1}\right)^2 = \frac{\left(1 + \sum_{i=2,3..} \left((M_i / M_1)(\omega_i / \omega_1)^2 (G_i / G_1)^2\right)\right)}{\left(1 + \sum_{i=2,3..} \left((M_i / M_1)^2 (G_i / G_1)^2\right)\right)} \quad [2.7]$$

Since $\omega_i \geq \omega_1$, the numerator cannot be less than the denominator. This means that $\omega \geq \omega_1$. Thus the calculated frequency will be an upperbound to the fundamental natural frequency. This is a very important result which paves the way for an improved procedure to find the natural frequencies as explained later.

It is important to stress here that the assumed displacement form must be *admissible*. In other words, it should be possible for the system to be given an initial displacement in the assumed form. This means, the assumed form must not violate any geometric constraint and must not introduce any geometric discontinuity within the structure. The set of admissible functions must also be complete, which means that it must contain all modes. For a beam problem, this implies that at rigid supports the displacement must be zero, and at built-in supports the slope must also be zero. In addition, the displacement and slope must be continuous within the length of the beam, and in the case of assembled structures the displacement and slope must also be continuous at any rigid connection point. Let us study the implication of Rayleigh's principle using an example.

2.3. Example: a simply supported beam

Consider the free vibration of a simply supported uniform thin beam having flexural rigidity EI, mass per unit length m and length L as shown in Figure 2.1.

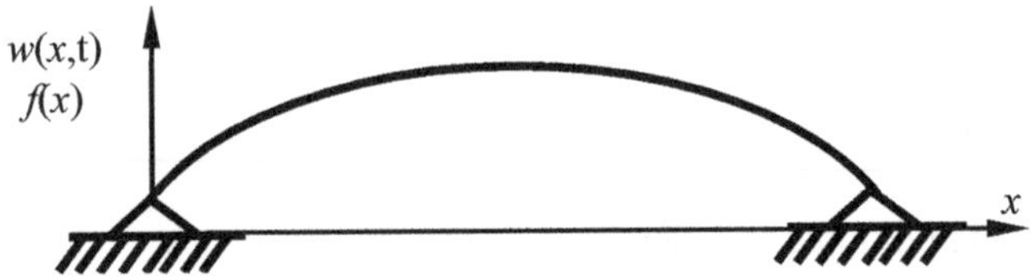

Figure 2.1. *A simply supported beam*

Let the lateral dynamic displacement be $w(x,t) = f(x)\sin(\omega t + \alpha)$.

The maximum total potential energy of the vibrating beam (see Chapter 6 for a list of potential and kinetic energy terms) is given by:

$$V_m = \int_0^L \frac{EI}{2}\left(f''(x)\right)^2 dx$$

The maximum velocity is $\omega f(x)$. Therefore the maximum kinetic energy due to vibration is:

$$T_m = \omega^2 \int_0^L \frac{m}{2}\left(f(x)\right)^2 dx$$

From Rayleigh's principle; $T_m = V_m$ gives an upperbound estimate of the fundamental natural frequency if an admissible function is used for $f(x)$.

$$\text{i.e. } \omega^2 \int_0^L \frac{m}{2}\left(f(x)\right)^2 dx = \int_0^L \frac{EI}{2}\left(f''(x)\right)^2 dx$$

Therefore

$$\omega^2 = \frac{\int_0^L \frac{EI}{2}\left(f''(x)\right)^2 dx}{\int_0^L \frac{m}{2}\left(f(x)\right)^2 dx} \qquad [2.8]$$

Using the exact fundamental mode function $\sin(\pi x / L)$ for $f(x)$ into the Rayleigh quotient gives:

$$\omega = (\pi / L)^2 \sqrt{(EI / m)} \qquad [2.9]$$

This is the exact value. Any other admissible function for f results in a higher value for the frequency. In this case the admissibility conditions are that the displacement must be zero at the two supports. i.e. $f(0) = 0$ and $f(L) = 0$.

$f(x) = G\,(x/L)(1-(x/L))$ is also admissible. Substituting this into equation [2.8] gives:

$$\omega = (10.95 / L^2)\sqrt{(EI / m)} \qquad [2.10]$$

This is about 11% higher than the exact value.

It is interesting to study the effect of using an over-constrained function such as $1-\cos(2\pi x / L)$. This function satisfies both geometric constraints. In addition, its slope is also zero at both supports. This means it would be an admissible function for a clamped-clamped beam. Using this function in equation [2.8] yields:

$$\omega = (22.79 / L^2)\sqrt{(EI / m)} \qquad [2.11]$$

This is 2% higher than the exact value of the fundamental natural frequency of a clamped – clamped beam. It is an upperbound estimate for the simply supported beam too, although it is a poor approximation.

From the above example, it is clear that the results obtained using Rayleigh's principle is dependent on the type of admissible function used. If the function used resembles the fundamental mode, the resulting estimate of the frequency is likely to be close to the exact fundamental natural frequency. It may also be noted that this approach gives only one value for the frequency. Therefore, for continuous systems, the higher natural frequencies cannot be obtained in this way. However, a more elaborate method, which may be regarded as an extension of Rayleigh's method, can be used to overcome some of these problems. This is called the Ritz method, or the Rayleigh–Ritz method. Before discussing this technique, let us study the meaning of admissibility through some examples.

2.4. Admissible functions: examples

The displacement forms in a Rayleigh–Ritz procedure must be continuous and satisfy all geometric constraints. Spatial functions that satisfy these conditions are called admissible functions. It is worth noting that the natural boundary conditions such as force = 0, or moment = 0 do not have to be satisfied. This is illustrated in the following examples.

1) Cantilever

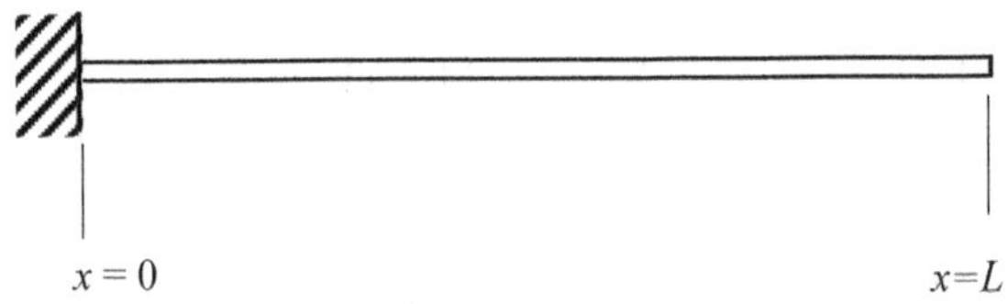

Figure 2.2. *Cantilever*

Geometric constraints:

At $x = 0$ $f = 0$

At $x = 0$ $df / dx = f' = 0$

Admissible functions:

$f(x) = (x / L)^i$ for $i = 2,3,4...$

2) Propped cantilever

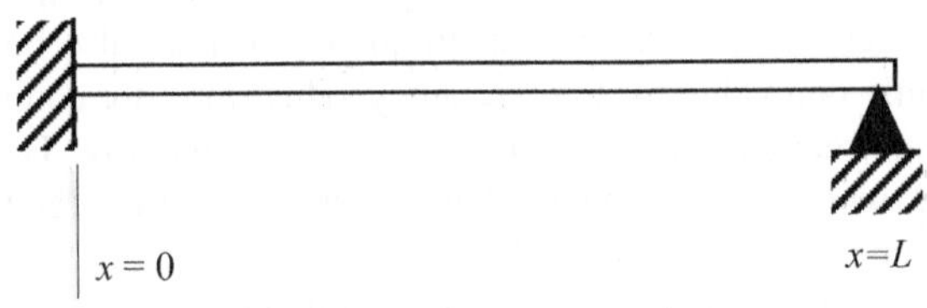

Figure 2.3. *Propped cantilever*

Geometric constraints:

At $x = 0$ $f = 0$

At $x = 0$ $df / dx = f' = 0$

At $x = L$ $f = 0$

Admissible functions:

$$f(x) = (x/L)^i (1-(x/L)) \text{ for } i = 2,3,4...$$

3) Pinned-pinned beam

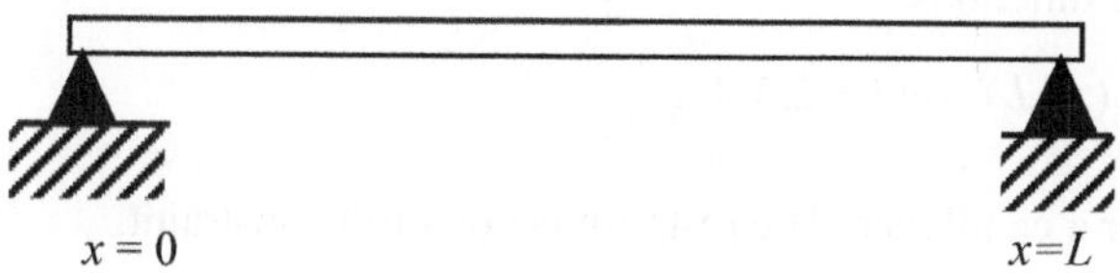

Figure 2.4. *Pinned-p*

Geometric constraints:

At $x = 0$ $f = 0$

At $x = L$ $f = 0$

Admissible functions:

$$f(x) = (x/L)^i (1-(x/L)) \text{ for } i = 1,2,3,4...$$

$$f(x) = \sin(i\pi x/L) \text{ for } i = 1,2,3,4...$$

4) Restrained cantilever

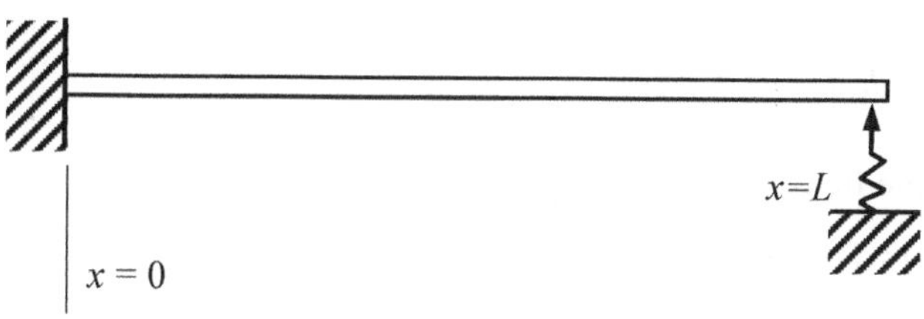

Figure 2.5. *Restrained cantilever*

Geometric constraints:

At $x = 0$ $f = 0$

At $x = 0$ $df/dx = f' = 0$

Admissible functions:

$f(x) = (x/L)^i$ for $i = 2,3,4...$

Same as for a cantilever. The restraint is not a full constraint.

5) Clamped-clamped

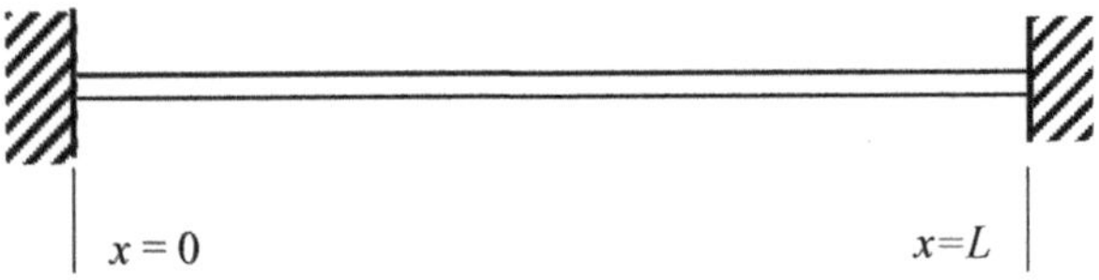

Figure 2.6. *Clamped-clamped*

Geometric constraints:

At $x = 0$ $f = 0$

At $x = 0$ $df/dx = f' = 0$

At $x = L$ $f = 0$

At $x = L$ $df/dx = f' = 0$

Admissible functions:

$f(x) = (1 - \cos(2i\pi x/L))$ for $i = 1,2,3,4...$

This set is also admissible for all other beams in this section but would result in an over constrained solution except for this case.

6) Pinned-free

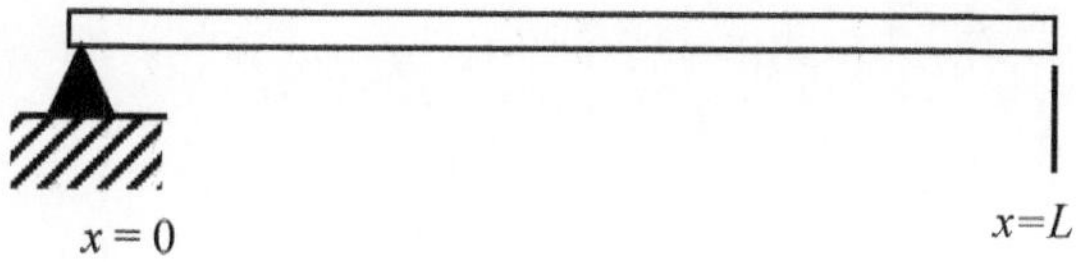

Figure 2.7. *Pinned-free*

Geometric constraints:

At $x = 0$ $f = 0$

Admissible functions:

$f(x) = (x / L)^i$ for $i = 1, 2, 3, 4...$

3

The Rayleigh–Ritz Method and Simple Applications

3.1. The Rayleigh–Ritz method

It was observed earlier that the natural frequency calculations based on the application of Rayleigh's principle are sensitive to the assumed displacement form, and that only one frequency can be determined. An extension of Rayleigh's principle, which also enables us to determine the higher frequencies, is the Rayleigh–Ritz method. We have seen that the lowest possible value of the frequency given by the Rayleigh quotient is an upperbound estimate of the fundamental natural frequency. Therefore, we can find the best estimate of the fundamental natural frequency by minimizing the Rayleigh quotient with respect to the undetermined mode shape. In this chapter, we will only show how to apply the method. More rigorous mathematical derivations that explain the theoretical basis of the method are given in Chapter 7.

In the example in section 1.3, minimizing the right hand side of equation [1.2] with respect to η gives $\eta = 0.75$ or 2. Substituting these back into the Rayleigh quotient yields the first and second natural frequencies, respectively. Since the Rayleigh quotient gives an upperbound to the fundamental natural frequency, the best possible estimate of the fundamental natural frequency is obtained by minimizing the Rayleigh quotient. This is the basis of the Rayleigh–Ritz method.

The Rayleigh–Ritz method is more commonly used in continuous systems where the maximum displacement f (say $f(x)$ for a one-dimensional continuous system) is expressed as the sum of a series of products of undetermined weighting coefficients and admissible displacement functions. The frequency expression (the Rayleigh quotient) is obtained in terms of the weighting coefficients. At this stage, the weighting coefficients are undetermined. Using Rayleigh's principle, the calculated frequency is an upperbound to the fundamental natural frequency. Therefore, the best possible estimate is obtained by minimizing the expression for the natural frequency with respect to the weighting coefficients. The resulting equations form a typical eigenvalue matrix equation. The roots are improved estimates of natural frequencies. The general derivation is given below:

$$f = \sum G_j \bar{\varphi}_j \qquad [3.1]$$

The maximum potential energy and maximum kinetic energy can be expressed in terms of the undetermined weighting coefficients G_j and the assumed displacement functions $\bar{\varphi}_j$.

Let $T_m = \omega^2\ \psi_m$, [3.2]

where ψ_m is a kinetic energy function.

Using Rayleigh's principle,

$$\omega^2 = V_m / \psi_m \qquad [3.3]$$

is an upperbound of ω_1^2.

To obtain the best possible estimate,

$\partial \omega_m^2 / \partial G_i = 0$ for, i=1,2.. [3.4]

Differentiating equation [3.3],

$$\frac{\partial \omega^2}{\partial G_i} = \frac{\psi_m(\partial V_m / \partial G_i) - V_m(\partial \psi_m / \partial G_i)}{\psi_m^2}$$

$$= \frac{(\partial V_m / \partial G_i) - (V_m / \psi_m)(\partial \psi_m / \partial G_i)}{\psi_m}$$

$$= \frac{(\partial V_m / \partial G_i) - \omega^2 (\partial \psi_m / \partial G_i)}{\psi_m},$$

using equation [3.3].

As ψ_m is finite, $\dfrac{\partial V_m}{\partial G_i} - \omega^2 \dfrac{\partial \psi_m}{\partial G_i} = 0$. [3.5]

If the number of terms in the series for f is n, there will be n such equations. For linear problems, each equation will be a linear function of G_is. Therefore, equation [3.5] may be written in a matrix form as follows:

$$[K]\{G\} - \omega^2 [M]\{G\} = \{0\} \qquad [3.6]$$

This is the generalized eigenproblem. The coefficients of the matrices $[K]$ and $[M]$ are found by using the following equations:

$$K_{i,j} = \frac{\partial^2 V_m}{\partial G_i \partial G_j} \qquad [3.7]$$

and

$$M_{i,j} = \frac{\partial^2 \psi_m}{\partial G_i \partial G_j} \qquad [3.8]$$

For continuous systems, the functions V_m and ψ_m usually involve integrals. The solution to the eigenvalue equation [3.6] gives n natural frequencies (eigenvalues) and corresponding modes (eigenvectors). From Rayleigh's theorem of separation, it can be shown that the frequencies calculated using this method are upperbound to the exact values as explained in section 7.1.

3.2. Application of the Rayleigh–Ritz method

a) Static and dynamic analysis of a cantilever

Dynamic analysis (finding natural frequencies)

Consider the lateral vibration of a cantilever beam of length L, mass per unit length m and flexural rigidity EI as shown in Figure 3.1.

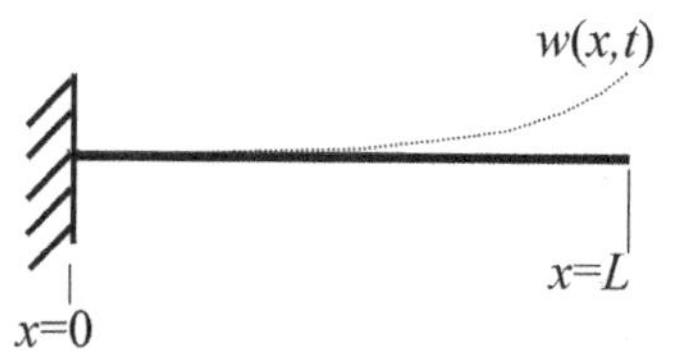

Figure 3.1. *Vibrating cantilever*

Let the lateral dynamic displacement be $w(x,t)=f(x)\sin(\omega t+\phi)$, where

$$f(x) = G_1 (x/L)^2 + G_2 (x/L)^3 \qquad [3.9]$$

Both functions $(x/L)^2$ and $(x/L)^3$ are admissible, since they are continuous for $0 \leq x \leq L$ and satisfy the geometric constraints that $f(0) = 0$ and $f'(0) = 0$.

$$V_m = \int_0^L \frac{EI}{2}\left(f''(x)\right)^2 dx$$

$$\Psi_m = \int_0^L \frac{m}{2}\left(f(x)\right)^2 dx$$

Substituting equation [3.9] into the above equations and carrying out the integration gives:

$$V_m = (EI/L^3)(2(G_1)^2+6\,G_1G_2+6(G_2)^2) \qquad [3.10]$$

$$\Psi_m = (mL/2)((1/5)(G_1)^2+(2/6)\,G_1G_2+(1/7)(G_2)^2) \qquad [3.11]$$

Now differentiating these energy terms with respect to G_1 and G_2 yields:

$$\partial V_m/\partial G_1 = (4EI/L^3)\,G_1 + (6EI/L^3)\,G_2 \qquad [3.12a]$$

$$\partial V_m/\partial G_2 = (6EI/L^3)\,G_1 + (12EI/L^3)\,G_2 \qquad [3.12b]$$

$$\partial \Psi_m/\partial G_1 = (mL/5)\,G_1 + (mL/6)\,G_2 \qquad [3.13a]$$

$$\partial \Psi_m/\partial G_2 = (mL/6)\,G_1 + (mL/7)\,G_2 \qquad [3.13b]$$

From the Rayleigh–Ritz procedure, the best possible estimate of the fundamental natural frequency is obtained by solving the following equations:

$$\partial V_m/\partial G_1 - \omega^2\,(\partial \Psi_m/\partial G_1) = 0 \qquad [3.14a]$$

$$\partial V_m/\partial G_2 - \omega^2\,(\partial \Psi_m/\partial G_2) = 0 \qquad [3.14b]$$

Substituting equations [3.12a], [3.12b] and [3.13a], [3.13b] into equations [3.14a], [3.14b] yields the following equation:

$$[K]\{G\} - \omega^2\,[M]\{G\} = \{0\}, \qquad [3.15]$$

where $[K]=\begin{bmatrix} 4EI/L^3 & 6EI/L^3 \\ 6EI/L^3 & 12EI/L^3 \end{bmatrix}$ and $[M]=\begin{bmatrix} mL/5 & mL/6 \\ mL/6 & mL/7 \end{bmatrix}$.

Equation [3.15] is a generalized eigenproblem and can be solved using any convenient method. There will be two roots for the frequency, giving the estimates of the first and second natural frequencies.

3.2.1.1. *Short cut to setting up the stiffness and mass matrices*

Generally, it is easier to perform the differentiation (minimization) with respect to the weighting coefficients first, followed by integration with respect to dx. For example, in the present case,

$$K_{1,2}=\frac{\partial^2 V_m}{\partial G_1 \partial G_2}=\frac{\partial^2}{\partial G_1 \partial G_2}\left(\int_0^L \frac{EI}{2}(f''(x))^2\,dx\right)=\int_0^L \frac{\partial^2}{\partial G_1 \partial G_2}\frac{EI}{2}(f''(x))^2\,dx$$

$$=\int_0^L \frac{\partial^2}{\partial G_1 \partial G_2}\sum_{i=1..}^{2}\frac{EI}{2}\left(G_1\left(\frac{2}{L^2}\right)+G_2\left(\frac{6x}{L^3}\right)\right)^2 dx = EI\int_0^L\left(\frac{12x}{L^5}\right)dx$$

$$=\frac{12EI}{L^5}\left(\frac{x^2}{2}\right)\Bigg|_0^L=\frac{6EI}{L^3}$$

Static analysis as a special case

If T_m = 0, then the Rayleigh–Ritz equations reduce to the equations corresponding to the minimization of total potential energy. The tip deflection of a cantilever due to a load P (see Figure 3.2) may be obtained as follows:

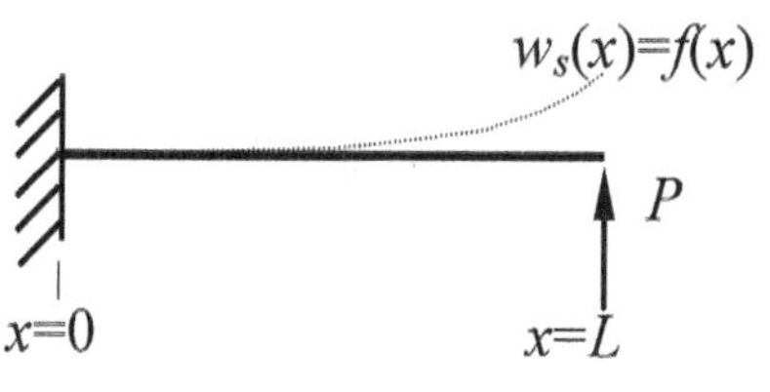

Figure 3.2. *Cantilever under static end load*

Additional potential energy due to $P=-Pf(L)=-P(G_1+G_2)$

Total potential energy $V_{sm} = (EI/L^3)(2(G_1)^2 + 6\, G_1\, G_2 + 6(G_2)^2) - P\,(G_1 + G_2)$

$$\partial V_{sm} / \partial G_1 = 4(EI / L^3)G_1 + 6(EI / L^3)G_2 - P = 0$$

$$\partial V_{sm} / \partial G_2 = 6(EI / L^3)G_1 + 12(EI / L^3)G_2 - P = 0$$

Solving the last two equations gives: $G_1 = PL^3/(2EI)$ and $G_2 = -PL^3/(6EI)$

$f(L) = G_1\,(L/L)^2 + G_2(L/L)^3 = G_1 + G_2 = PL^3/(3EI)$

This is the exact answer for the tip deflection of a cantilever. This example shows that the Rayleigh–Ritz approach can also be used for static analysis.

b) Vibration of a partially restrained cantilever beam carrying a mass

Let us now modify the problem and consider the vibration of a cantilever beam carrying a thin mass and partly restrained by a spring at a point. The mass and moment of inertia of the thin mass are m_0 and I_0, respectively. The spring constant is k. They are attached as shown in Figure 3.3.

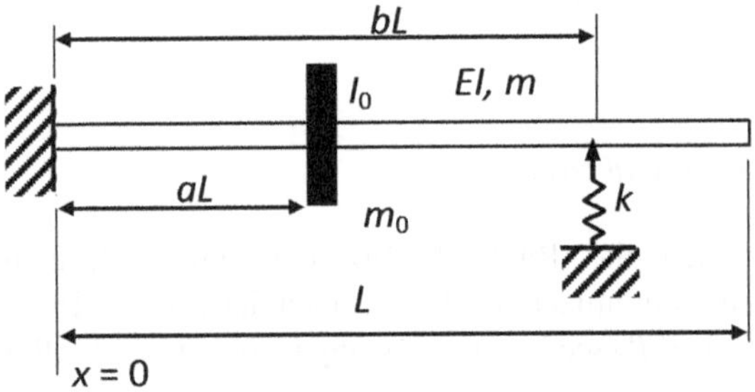

Figure 3.3. *Restrained cantilever with a rigid mass*

Let us choose a series

$$f(x) = \sum_{j=1,2}^{n} G_j \bar{\varphi}_j \quad \text{where } \bar{\varphi}_j = (x / L)^{j+1} \qquad [3.16]$$

The maximum total potential energy includes potential energy due to the bending of the beam itself and the stretching or shortening of the attached spring. Similarly, the maximum total kinetic energy of the system includes kinetic energy due to the vibratory motion of the beam, and the lateral motion and rotation of the thin mass.

$$V_m = V_{beam} + V_{spring}$$

$$\psi_m = \psi_{beam} + \psi_{mass} + \psi_{rotation}$$

The potential energy due to the spring is:

$$V_{spring} = (k/2)(f(bL))^2$$

The kinetic energy due to the thin mass is:

$$\psi_{mass} + \psi_{rotation} = (m_0/2)(f(aL))^2 + (I_0/2)(f'(aL))^2$$

Combining the above expressions, the maximum total potential energy of the system is:

$$V_m = \int_0^L (EI/2)(f''(x))^2 dx + (k/2)(f(bL))^2$$

The maximum total kinetic energy function is:

$$\psi_m = \int_0^L (m/2)(f(x))^2\, dx + (m_0/2)(f(aL))^2 + (I_0/2)(f'(aL))^2$$

$$K_{i,j} = \frac{\partial^2}{\partial G_i G_j}\left(\int_0^L (EI/2)(f''(x))^2 dx + (k/2)(f(bL))^2\right)$$

$$= \frac{\partial^2}{\partial G_i G_j}\left(\left((EI/2)\int_0^L\left(\sum_{j=1,2..}^{n} G_j \bar{\varphi}_j''(x)\right)^2 dx\right) + (k/2)\left(\sum_{j=1,2..}^{n} G_j \bar{\varphi}_j(bL)\right)^2\right)$$

$$= EI\int_0^L \bar{\varphi}_i''\bar{\varphi}_j'' dx + k\bar{\varphi}_i(bL)\bar{\varphi}_j(bL) \qquad [3.17]$$

Substituting equation [3.16] into the above gives:

$$K_{i,j} = \int_0^L \left(EI(i+1)i\left(\frac{x}{L^{i+1}}^{i-1}\right)(j+1)j\left(\frac{x}{L^{j+1}}^{j-1}\right)\right) dx + k(b)^{i+1}(b)^{j+1} =$$

$$\int_0^L EI(i+1)i(j+1)j\left(\frac{x^{i+j-2}}{L^{i+j+2}}\right) dx + kb^{i+j+2} = \frac{(i+1)i(j+1)jEI}{(i+j-1)L^3} + kb^{i+j+2} \qquad [3.18]$$

Let us now find the mass matrix.

$$M_{i,j} = \frac{\partial^2}{\partial G_i G_j}\left(\int_0^L (m/2)(f(x))^2\, dx + (m_0/2)(f(aL))^2 + (I_0/2)(f'(aL))^2 \right)$$

This gives: $M_{i,j} = m\int_0^L \overline{\varphi}_i\overline{\varphi}_j dx + m_0\overline{\varphi}_i(aL)\overline{\varphi}_j(aL) + I_0\overline{\varphi}_i'(aL)\overline{\varphi}_j'(aL)$

Substituting equation [3.16] into the above we get:

$$M_{i,j} = \frac{mL}{(i+j+3)} + m_0 a^{i+j+2} + I_0(i+1)(j+1)a^{i+j} \qquad [3.19]$$

As there are two discontinuities in the beam, introducing three segments, to solve this problem using the exact method would require setting up a 12 × 12 determinantal formulation requiring substantial analytical work. With the Rayleigh–Ritz approach, the above equations could be solved using any standard eigenvalue solver. The solution to the above problem is available as an interactive program asymptot.exe available online from the author's website [ILA 00a].

The program gives that natural frequency parameter λ defined by

$$\lambda = L\left(\frac{m\omega^2}{EI}\right)^{1/4}$$

The mass and/or spring may be attached to any point on the beam. The program does not take into account of the rotary inertia of the beam itself but includes the kinetic energy due to the flexural rotation of the added mass if its moment of inertia is specified.

It is possible to obtain the natural frequencies of clamped-simply supported beams by including a spring with a very large stiffness or a very large mass at the free end of the cantilever beam. The functions used in the program are only admissible for the cantilever beam, but in the limiting case, a large restraint in the form of a spring or a concentrated mass would effectively prevent the translation at the point where it is located. In this way, we may be able to overcome the limitations on the choice of admissible functions (see Chapters 5 and 7). By adding a very large mass with a very large moment of inertia, it is possible to effectively clamp the beam at the point where it carries the mass, and obtain results for a clamped-clamped beam.

The program asymptot.exe [ILA 00a] also contains a module to calculate the frequencies of a similar system where the clamp at the left end has been replaced by a simple support (pinned-free beam). We can also obtain the frequencies of a pinned-pinned beam, by attaching a large mass (without any moment of inertia) to the free end of the pinned-free beam. However, it should be noted that for higher modes, the magnitude of mass or spring stiffness that is required to prevent the translation (or rotation in the case of moment of inertia) may be too high, and using very large parameters in calculations can cause numerical problems. Another feature worth noting here is that when using very large masses or moments of inertia, there may be one or two modes at very low frequencies, when the system behaves like a discrete one. This is because the mass of the beam is negligible compared to that of the attached mass/inertia. These frequencies may be calculated using an approximate discrete system model in which the structure provides the stiffness only. In these modes, translation or rotation of the mass does take place. Unless a sufficiently small value for the search increment dλ is used, these low frequency modes may be missed. As an exercise, find the natural frequencies and the modes using the interactive program for simple boundary conditions and compare the results with exact values from the Newtonian method [ILA 00b].

Here are some results for the natural frequencies of a cantilever (without springs or attached masses) obtained using a MATLAB code in appendix A.1. Corresponding exact results are, using newtonian.exe available online at [ILA 00b], also given for comparison. The modes for the 9 term solution are shown in Figure 3.4.

No of terms	First mode	Second mode	Third mode	Fourth mode
1	2.115	–	–	–
2	1.88	5.9	–	–
3	1.875	4.715	10.869	–
4	1.875	4.707	7.959	16.781
5	1.875	4.694	7.952	11.337
6	1.875	4.694	7.856	11.331
7	1.875	4.694	7.856	11.005
8	1.875	4.694	7.855	11.005
9	1.875	4.694	7.855	10.996
Exact	1.875	4.694	7.855	10.996

Table 3.1. *Rayleigh–Ritz results and exact results for the first four modes of a cantilever*

From Table 3.1 it may be seen that with only two terms, the first natural frequency parameter can be calculated to within 1% of the exact value. For the second natural frequency, it takes 3 terms to reach a similar degree of accuracy. As a rule of thumb, we should use about twice as many terms as the number of modes of interest. However, we should be particularly careful when a system has attached masses or springs because if the magnitude of the spring stiffness or the mass was large then the corresponding displacement or rotation is forced to become zero, which implies that the coefficients G_is are forced to satisfy the corresponding constraint condition. This then effectively removes the independence of these coefficients. For example, a two term approximation with a very stiff spring at $x = bL$ would make $f(bL)$ zero. This means $G_1 b^2 + G_2 b^3 = 0$ giving $G_1 = - G_2 b$. This is equivalent to using a function $f(x) = G_2\left(\left(\frac{x}{L}\right)^3 - b\left(\frac{x}{L}\right)^2\right)$ which amounts to a single term approximation. Therefore, we should add one extra term for each degree of freedom that is likely to be lost due to the inclusion of a mass, inertia or stiffness. If a body possessing both inertia and mass is included, then two additional terms are recommended.

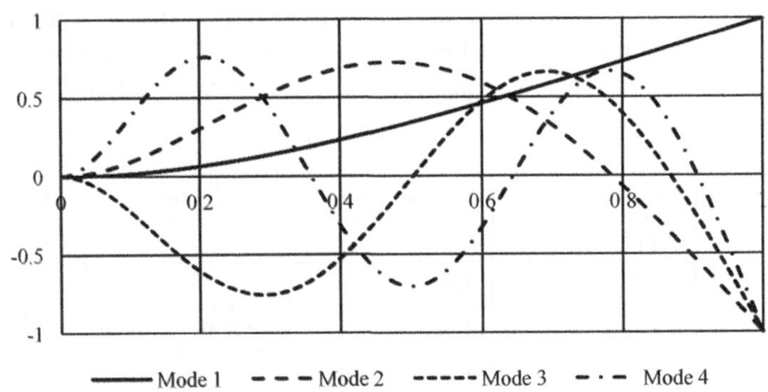

Figure 3.4. *Vibration modes of a cantilever*

In deciding on the number of terms it should be remembered that depending on the choice of functions, large number of terms could cause computational (numerical) problems. Therefore, a numerical convergence study is always recommended to see if the solution is reliable. For example using the MATLAB program in appendix A.1, which is based on the polynomial functions, we can see that the frequency parameters converge well up to $n = 12$ but for $n = 13$, they suddenly deviate. This is due to the behavior of the higher order polynomial terms which could lead to large round-off errors and ill-conditioning. This means to obtain accurate estimates of the natural frequencies of higher modes, we have to look for other admissible functions. The choice of admissible functions is the subject of Chapter 8. For now, we will continue to use the polynomial functions for the purpose of understanding how to handle different geometric boundary conditions.

4

Lagrangian Multiplier Method

4.1. Handling constraints

The difficulty in choosing admissible functions has been the most significant drawback of the Rayleigh–Ritz method. One way to relax the admissibility requirement may be found by utilizing a constrained optimization technique dating back to the 18th Century. That is by using the Lagrangian multiplier method [LAG 01, FRA 05], in conjunction with the Rayleigh–Ritz method. This means individual functions need not satisfy the geometric boundary conditions but the series as a whole is forced to satisfy these by including additional constraint equations. To explain this point we could consider the following metaphor. The manager of a marketing company may tell its sales agents that the company needs to meet a certain sales target within six months and that each agent must meet a certain sales target every week. While setting a target may be desirable for a company, prescribing fine details such as weekly targets could place severe constraints on its employees. Instead, if they are instructed to reach a certain target over six months, the additional flexibility for the individuals gives them the choice to better plan their own efforts. It is the final outcome that is important and not whether the targets are achieved every week. To illustrate how the Lagrangian multiplier method works, let us formulate the procedure for calculating the frequencies of a propped cantilever [ILA 99].

4.2. Application to vibration of a constrained cantilever

Let the lateral dynamic displacement be $v(x,t)=f(x)\sin(\omega t+\alpha)$, where:

$$f(x) = \sum_{j=1,2..}^{n} G_i \left(\frac{x}{L}\right)^{j+1} \qquad [4.1]$$

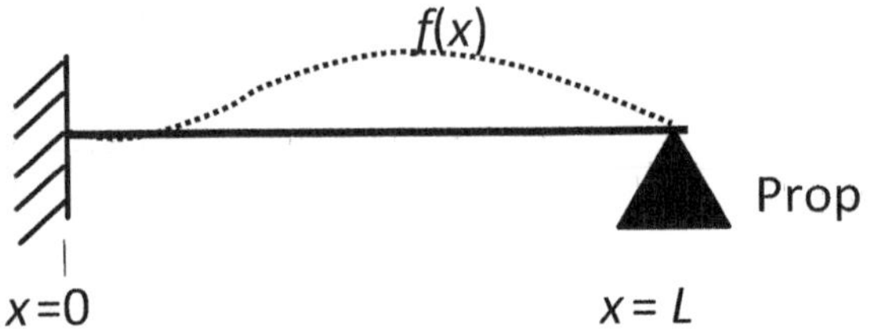

Figure 4.1. *Propped cantilever*

The functions $(x/L)^2$ and $(x/L)^3$ are not admissible although they are continuous for $0 \le x \le L$ and satisfy two of the geometric constraints that $f(0) = 0$ and $f'(0) = 0$ because they fail to satisfy the condition that $f(L) = 0$. However we can make the sum of the series satisfy this condition by incorporating the constraint equation that:

$$f(L) = \sum_{j=1,2..}^{n} G_j = 0 \qquad [4.2]$$

In the Lagrangian multiplier method, the minimization equations are also modified as follows.

$$\frac{\partial V_m}{\partial G_i} - \omega^2 \frac{\partial \psi_m}{\partial G_i} + \frac{\partial}{\partial G_i}\left(G_{n+1} f(L)\right) = 0 \qquad [4.3]$$

G_{n+1} is a constant called a Lagrangian multiplier. In the present example there is only one constraint condition to be satisfied namely displacement *at* $x=L$ is zero given by equation [4.2] and hence only one Lagrangian multiplier. If we had two props or a clamped support at the right end which would have prevented translation as well as rotation then we would have had two constraint equations and two multipliers, say G_{n+1} and G_{n+2}.

The constraint equation is taken as the $(n+1)$th equation.

The potential and kinetic energy terms are given by:

$$V_m = \int_0^L \frac{EI}{2}\left(f''(x)\right)^2 dx \qquad [4.4]$$

and

$$\psi_m = \int_0^L \frac{m}{2}\left(f(x)\right)^2 dx \qquad [4.5]$$

Substituting equations [4.1], [4.2], [4.4], [4.5] into equation [4.3] results in the matrix equation:

$$[K]\{G\} - \omega^2 [M]\{G\} = \{0\} \qquad [4.6]$$

where both $[K]$ and $[M]$ are of dimensions $(n+1)\times(n+1)$ and vector $\{G\}$ has a dimension $(n+1)$.

All terms in the first n rows and column are the same as that for the typical Rayleigh–Ritz equation. That is, for $i < (n+1)$, and $j < (n+1)$,

$$K_{i,j} = \frac{(i+1)i(j+1)jEI}{(i+j-1)L^3} \text{ and } M_{i,j} = \frac{mL}{(i+j+3)}$$

The elements of the last row and last column are however different and are given by the following equations:

For $j < n+1$, from equation [4.3], $K_{i,j} = 1$ and $M_{i,j} = 0$

Since equation [4.2] is our $(n+1)$th equation, for $i < n+1$, $K_{i,j} = 1$ and $M_{i,j} = 0$.

Also, note that $K_{n+1,n+1} = 0$ and $M_{n+1,n+1} = 0$.

This is not a typical eigenvalue equation, as the last term in the main diagonal of both matrices is zero, but it can be solved. The first four frequency parameters calculated, using up to five terms in the series, are compared with exact results for a clamped-simply supported beam in Table 4.1. These were determined using the MATLAB code in Appendix A.2. The modes are given in Figure 4.2.

No of terms n	λ_1	λ_2	λ_3	λ_4
1	-	-	-	
2	4.5270	-	-	-
3	3.9308	8.6794	-	-
4	3.9285	7.1221	13.461	-
5	3.9266	7.0985	10.428	18.999
6	3.9266	7.0696	10.351	13.927
7	3.9266	7.0688	10.218	13.752
8	3.9266	7.0686	10.213	13.385
9	3.9266	7.0686	10.210	13.370
Exact	3.9266	7.0686	10.210	13.352

Table 4.1. *Frequency parameters obtained using the Lagrangian multiplier method*

It should be noted here that compared to the standard Rayleigh–Ritz method where individual functions satisfy all geometric constraint conditions, in the Lagrangian multiplier method, we need to choose extra terms in the series, equal in number to the constraints to be introduced. In the present case, we are modeling an extra pin support. Thus it is necessary to choose at least one more term because of the need to satisfy equation [4.2].

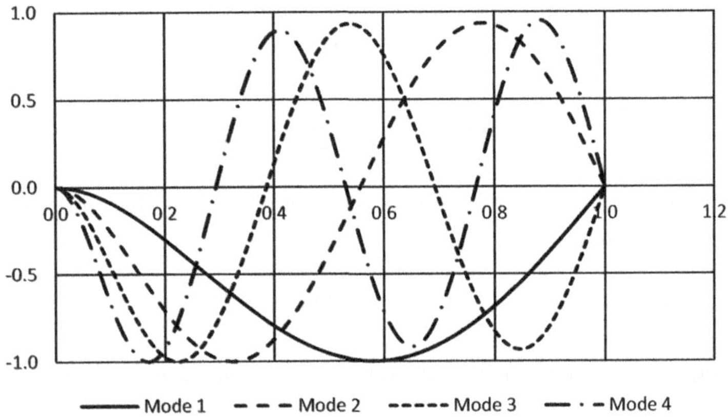

Figure 4.2. *Vibration modes of a propped cantilever*

There are two problems with the Lagrangian multiplier method. One is that it requires a different formulation of the stiffness and mass matrices with the addition of each constraint. It is not possible to have one expression for the stiffness (or mass) matrix element that is applicable for the whole matrix. Another problem involves continuous constraints. While it is possible to write explicit expressions for the constraints when they involve a variable at a point, it is not possible to constrain a variable along a line or within an area exactly. For example if the deflection of a plate $f(x,y)$ is fully restrained along a line defined $ax+by = c$, the constraint equation becomes a function of x and y which makes it difficult to solve.

For these reasons, an asymptotic modeling approach discussed in the next chapter is often favored by researchers (see [ILA 99] for a list of references). To see the asymptotic modeling in an interesting way, let us consider the marketing strategy analogy in section 4.1 again. Let us ask the question, what is the best way to motivate the sales agents and get the optimum outcome? Giving them rigid targets or telling them that they would get a bonus depending on how close they are to the target? If the agents realize that even if they do not reach the target they would still be rewarded for getting close to it, perhaps they will be better motivated. It should be pointed out here that the authors are not experts in marketing or management strategies, and these points are being mentioned to make this reading somewhat entertaining.

5

Courant's Penalty Method Including Negative Stiffness and Mass Terms

5.1. Background

So what is a better way to handle constraints? The answer came from a famous mathematician of the 19th/20th Century called Richard Courant. In his address to the American Mathematical Society [COU 43] Courant describes the difficulty in choosing "the coordinate functions" as the "more annoying" of the two "objections" to the Rayleigh–Ritz method, the less annoying one being "the vagueness as to the accuracy of the approximation". This difficulty arises mainly because of the need to satisfy the geometric constraints of a system. In the same reference, Courant suggests the use of artificial stiffness parameters having very large magnitude so that a rigid constraint could be approximately modeled by a restraint. The effectiveness of this approach has since been studied by several researchers for various interesting problems and its applicability has also been extended to analyze rigidly connected systems and systems with cracks (see [GOR 89, KIM 90, YUA 92a, YUA 92b, CHE 92, LEE 94]). This method has now evolved into what is commonly called the "penalty method". The high stiffness of the spring is effectively a penalty against violation of the constraint. The higher the stiffness, which may be referred to as a penalty parameter, the harder it is to violate the constraint. This method is used in Finite Element Analysis [ZIE 74, ZIE 77] and in many disciplines for solving constrained optimization or variational problems [AVI 98, GAV 00, ILA 02a, ILA 2b]. In more general terms, penalty function method involves the addition of quadratic penalty terms to the functional to be minimized. These additional terms are the product of the squares of error functions (error referring to the results of not satisfying constraints, for example in the propped cantilever case $f(L)$ is the error term) and large penalty values. If the value of the penalty parameter is very large then the error functions must remain very small, and in the limiting case as the penalty term approaches infinity the error should approach zero, because the product itself is a definite term. Let us illustrate this method by considering the propped cantilever example again.

5.2. Penalty method for vibration analysis

The idea is to replace the prop with a spring support of very high stiffness. The additional potential energy terms associated with the spring has already been considered in a previous example in section 3.2. The results for this case can be obtained by using the equations in section 3.2. Numerical experimentation shows that as the stiffness parameter is increased the natural frequencies increase but for very high values of stiffness they asymptotically approach the frequencies of the propped cantilever from below. In subsequent sections we will refer to such approximate models of constrained problems as asymptotic models or penalized models.

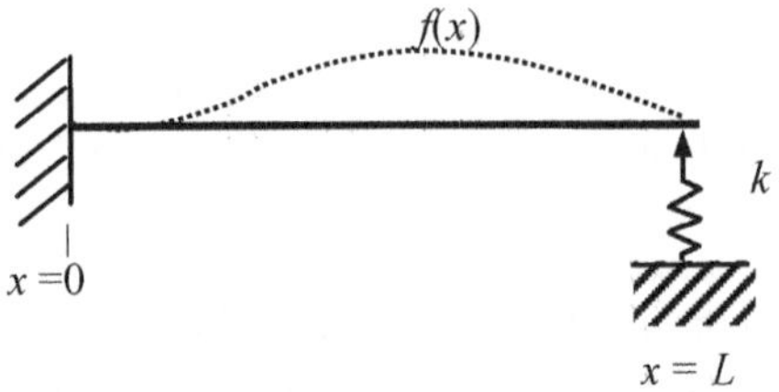

Figure 5.1 *Cantilever with an added partial restraint*

The results for the first three modes are given in Table 5.1. Moving along the columns, we can see that for any given penalty stiffness, the frequencies decrease with number of terms (admissible functions) in the series. The penalty effect is seen by moving along the rows. It can be seen that for any given number of admissible functions, with increasing penalty stiffness, the frequencies increase, as we would expect and can never decrease. The stiffness cannot of course exceed infinity – that is an ideally rigid support condition. This means the frequencies converge toward the frequencies of the constrained system (propped cantilever). So to use this method successfully we need to calculate the frequencies for increasing values of stiffness until a convergence is observed. It should also be borne in mind that in computing there is a practical limit beyond which we cannot increase the numerical values due to large round of errors and associated risk of ill-conditioning. So it would be ideal to know what stiffness values should be used. There are two problems with this method. One is that it is difficult to guess what magnitude of stiffness would give accurate results for the constrained system and even if convergence is observed for one frequency, the magnitude of stiffness may not be large enough to fully restrain the motion at the prop for higher modes. The second problem is that it gives an upper bound to the asymptotic model, which is somewhat flexible compared to the fully restrained model, and the frequency could have been lowered by the flexibility of the support. That is to say it gives an upper bound result to a model that effectively gives a lower bound to the original (constrained) system. This is best illustrated through a plot of the natural frequency parameter with penalty stiffness.

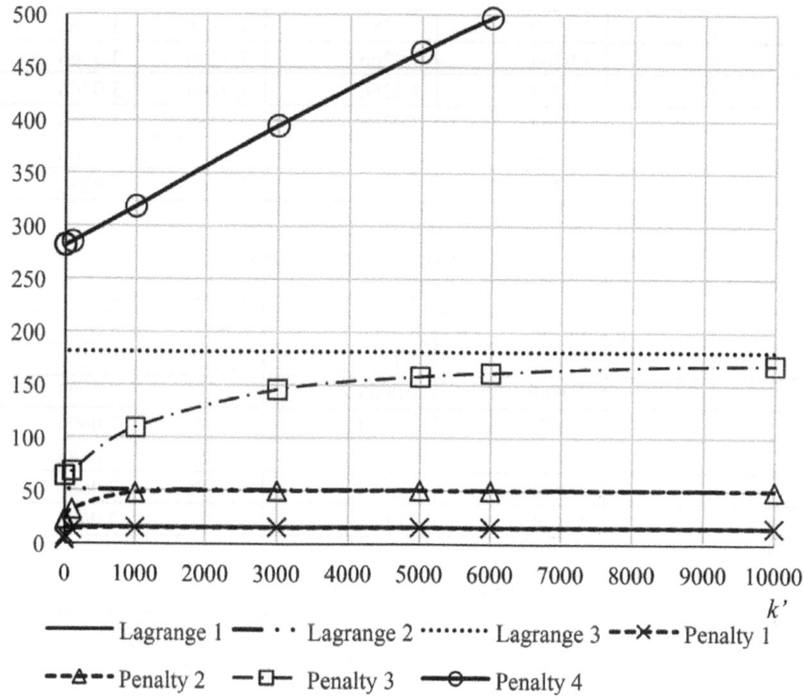

Figure 5.2. *Four term solution for a cantilever with spring support*

The non-dimensional frequency parameter λ^2 is plotted against non-dimensional penalty stiffness for a four term solution in Figure 5.2. The first three non-dimensional frequencies can be seen to approach the corresponding non-dimensional natural frequencies of a propped cantilever obtained using the

Lagrangian multiplier method. However, the highest frequency continues to increase monotonically and "goes through the roof". This is a feature of the penalty method. In this example we have only one constraint which is modeled by a restraint

First mode: exact λ_1=3.92660				
n	$\|\|k'\|\|$			LAGRANGIAN
	10^5	10^6	10^7	
2	4.5263	4.5269	4.5270	4.5270
3	3.9305	3.9307	3.9308	3.9308
4	3.9282	3.9285	3.9285	3.9285
5	3.9264	3.9266	3.9266	3.9266
6	3.9263	3.9266	3.9266	3.9266
7	3.9263	3.9266	3.9266	3.9266
8	3.9263	3.9266	3.9266	3.9266
9	3.9263	3.9266	3.9266	3.9266
Second mode: exact λ_2=7.0686				
2	33.103*	58.858*	104.66*	-
3	8.6712	8.6786	8.6793	8.6794
4	7.1202	7.1219	7.1220	7.1221
5	7.0966	7.0983	7.0985	7.0985
6	7.0678	7.0694	7.0696	7.0696
7	7.0671	7.0687	7.0688	7.0688
8	7.0668	7.0684	7.0686	7.0686
9	7.0668	7.0684	7.0686	7.0686
Third mode: exact λ_3=10.210				
2	-	-	-	-
3	38.106*	67.702*	120.38*	-
4	13.414	13.456	13.460	13.461
5	10.422	10.428	10.428	10.428
6	10.344	10.350	10.351	10.351
7	10.213	10.217	10.218	10.218
8	10.208	10.213	10.213	10.213
9	10.205	10.210	10.210	10.210

*These frequencies increase monotonically and the modes violate constraints.

Table 5.1. *Comparison of natural frequency parameters for the clamped-restrained beam*

If we had h number of constraints that are replaced by springs of high stiffness, then we can expect h number of such frequencies to monotonically increase. It is the remaining frequencies that approach the frequencies of the constrained system. Another drawback of the method is that the natural frequencies of the constrained system are approached from below. That is, the penalized solution is a lower bound to the fully constrained solution. When using the Rayleigh–Ritz method which gives an upper bound with respect to the number of terms being used, we then get an upper bound solution to a model which gives a lower bound. Thus the "boundedness" of the Rayleigh–Ritz solution is lost. However, there is a way to address this problem which was accidentally discovered through a careless mistake made by the Author of this chapter in his PhD thesis.

5.3. Penalty method with negative stiffness

We said there is a way to address these problems. In fact there are several ways! We will deal with one of them first. Recent work has shown that it is possible to use not only large magnitudes of positive stiffness but also large magnitudes of negative stiffness, to asymptotically model a rigid constraint. The reason is rather simple. Whether the sign is positive or negative, a stiffness of very large magnitude causes the associated displacement to become zero or very small by serving as a penalty. The good news is the way a positively penalized system approaches the constrained system is opposite to the way a negatively constrained system approaches it. For example, increasing the magnitude of a negative stiffness causes the frequencies to decrease and yet approach the frequencies of a constrained system at very large magnitudes, but from the opposite side! So by using negative stiffness of large magnitudes, we can get a true upper bound solution. Furthermore, by using a combination of positive and negative values for the spring stiffness, we can find the maximum possible error due to the asymptotic modeling by bracketing the constrained solution from opposite sides. This may be seen from the numerical results presented in Table 5.2 and Figure 5.3.

The results generated from the program asymptot.exe [ILA 00a] for various values of a non-dimensional stiffness parameter k' is presented in Table 5.2 together with exact values and frequencies obtained using the Lagrangian multiplier method. The non-dimensional stiffness parameter is defined as:

$$k' = \frac{kL^3}{EI}$$

It may be noted here that the constrained solution (i.e. obtained from Lagrangian multiplier method) is bracketed by results obtained using large positive and negative stiffness parameters.

First mode: exact λ_1=3.92660					
n	Sign of k'	$\lVert k' \rVert$			LAGRANGIAN
		10^5	10^6	10^7	
2	+	4.5263	4.5269	4.5270	4.5270
	-	4.5277	4.5271	4.5270	
3	+	3.9305	3.9307	3.9308	3.9308
	-	3.9310	3.9308	3.9308	
4	+	3.9282	3.9285	3.9285	3.9285
	-	3.9288	3.9285	3.9285	
5	+	3.9264	3.9266	3.9266	3.9266
	-	3.9269	3.9267	3.9266	
9	+	3.9263	3.9266	3.9266	3.9266
	-	3.9269	3.9266	3.9266	
Second mode: exact λ_2=7.0686					
2	+	33.103*	58.858*	104.66	-
	-	-	-	-	
3	+	8.6712	8.6786	8.6793	8.6794
	-	8.6875	8.6802	8.6795	
4	+	7.1202	7.1219	7.1220	7.1221
	-	7.1239	7.1222	7.1221	
5	+	7.0966	7.0983	7.0985	7.0985
	-	7.1004	7.0987	7.0986	
9	+	7.0668	7.0684	7.0686	7.0686
	-	7.0704	7.0688	7.0686	
Third mode: exact λ_3=10.210					
2	+	-	-	-	-
	-	-	-	-	
3	+	38.106*	67.702*	120.38	-
	-	-	-	-	
4	+	13.414	13.456	13.460	13.461
	-	13.507	13.465	13.461	
5	+	10.422	10.428	10.428	10.428
	-	10.434	10.429	10.428	
9	+	10.205	10.210	10.210	10.210
	-	10.216	10.211	10.210	

*These frequencies increase monotonically and the modes violate constraints

Table 5.2. *Comparison of natural frequency parameters for the clamped-restrained beam*

The variation of the non-dimensional natural frequency parameter λ^2 with added stiffness of positive and negative signs is shown in Figure 5.2. The figure shows that the frequencies increase with positive stiffness and decrease with increasing magnitude of negative stiffness. However, by the theorem of separation these increases and decreases cannot cross the natural frequencies of the constrained system, because which is asymptotically approaches as the magnitude of stiffness approaches infinity.

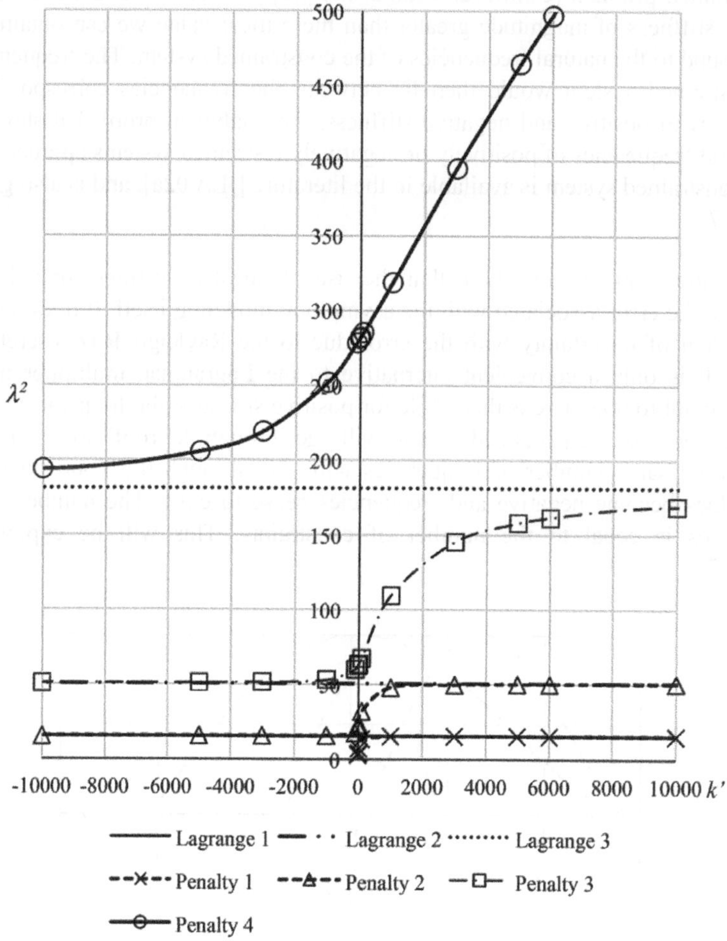

Figure 5.3. *Variation of the frequency parameter λ^2 with added stiffness*

An interesting point to note is that when the spring stiffness k' is equal to –3 natural frequency for one of the modes disappears. This is to be expected because the lateral stiffness of the cantilever at the tip is 3 kL^3 / (EI). Therefore when k' = -3 the system has no net stiffness and one of the frequencies become imaginary for negative stiffness having any larger magnitude than this. However this is of no concern because the second natural frequency of the system now becomes the fundamental and is the one that asymptotically approaches the fundamental natural frequency of the constrained system from above. It is not difficult to determine the critical value of the spring stiffness that is equal and opposite to the stiffness of the unconstrained problem (cantilever without the prop). Once this is done, by using a negative stiffness of magnitude greater than the critical value we can obtain a true upper bound to the natural frequencies of the constrained system. The frequencies of the constrained system would then lie between the frequencies corresponding to restraints with positive and negative stiffness. An analytical proof that shows that the natural frequencies of positively or negatively restrained systems approach those of the constrained system is available in the literature [ILA 02a], and is also given in Chapter 7.

It is important to stress here that the use of negative stiffness only helps to determine the error associated with the asymptotic modeling itself. It does not solve the problem of uncertainty with the error due to the Rayleigh–Ritz discretization process. It is only a convenient alternative to the Lagrangian multiplier method. Another point to note here is that while for positive stiffness, the frequencies of one or more modes of the penalized system will "go through the roof" for the negative stiffness the same number of frequencies will "fall through the floor" in that the eigenvalues become negative and frequencies cease to exist. The number of such frequencies is equal to the number of constraints. This will be explained in Chapter 7.

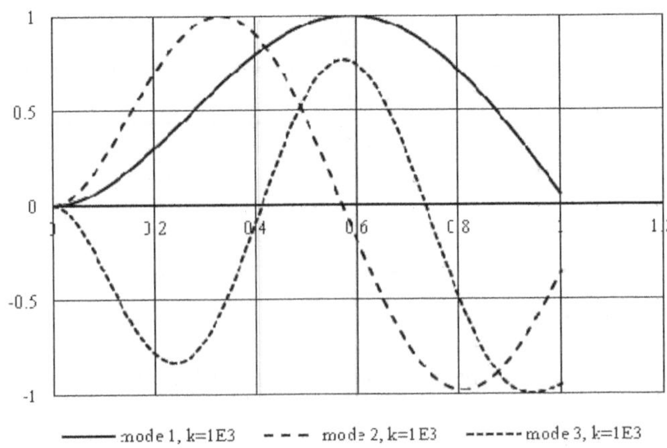

Figure 5.4. *First three modes of vibration for k=1000*

It is interesting to see the change in modes with penalty parameter. For a non-dimensional penalty stiffness of 1,000, the first three modes are shown in Figure 5.4. It may be noted that while the first mode has near zero displacement at the prop, the constraint violation increases with mode number. Figure 5.5 shows the modes for a non-dimensional stiffness of 10^4. Now all three modes show very little violation of the constraints. The modes have been normalized by making the magnitude of the maximum displacement unity and plotted against non-dimensional axial length between zero and one.

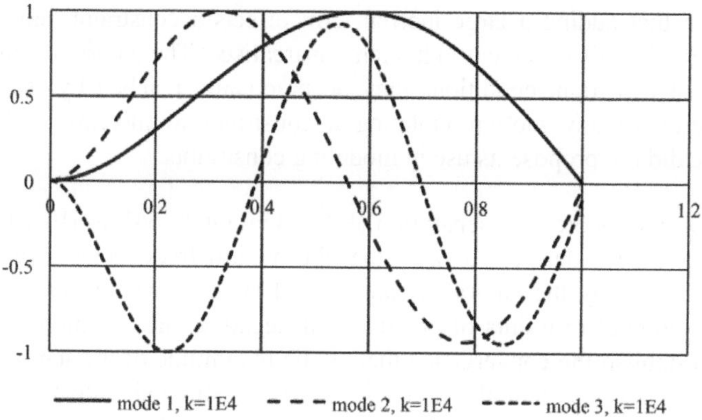

Figure 5.5. *First three modes of vibration for k=10^4*

5.4. Inertial penalty and eigenpenalty methods

Earlier we mentioned there being several ways to address the problems with the penalty method and here we present the second. Although the use of "artificial springs" of very high stiffness is the typical penalty method of handling constraints in vibration analysis, use of large mass (translational inertia) or moment of inertia (rotational inertia) could also serve the same purpose. It is well known that while increasing the stiffness may cause an increase in natural frequencies but not a decrease, with inertia the opposite is true. Thus it is possible to get a true upper bound for the natural frequencies of a constrained system using an asymptotic model where the constrained are replaced with inertial type penalty restraints [ILA 05b]. As the inertial penalty parameter such as the mass or moment of inertia are increased, the natural frequencies would tend to approach those of the constrained system but from below. This is good news because this is how the frequencies converge with respect to number of terms in the Rayleigh–Ritz method. Therefore we can get a true upper bound solution with an asymptotic (or penalized) model. Those who prefer to avoid using negative stiffness values may wish to take this route. The detailed proof is given in [ILA 05b].

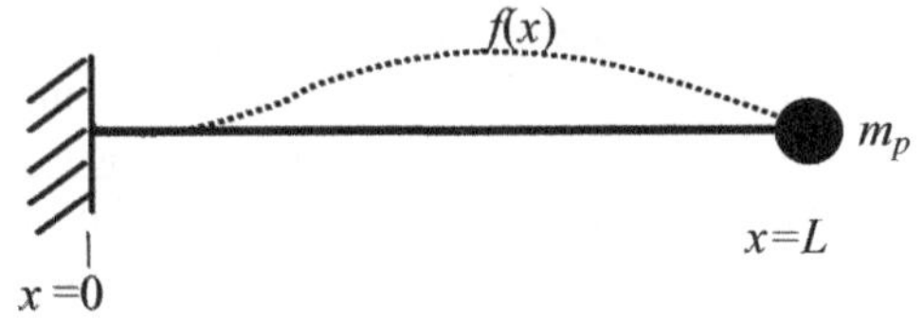

Figure 5.6. *Cantilever with a penalty mass*

The idea that adding a large inertial term effects a constraint may be inferred from Lord Rayleigh's statement in his famous treatise, "The theory of sound", where he states that constraint conditions may be practically reached by "supposing the kinetic energy of any motion violating a constraint to increase without limit", although he did not propose its use in modeling constraints.

Based on the theorem of separation [RAY 45a, GOU 66], in [ILA 05b] several theorems have been derived to show that the natural frequencies of a constrained system are bound by the natural frequencies of systems where the constraints are replaced by inertial restraints of positive and negative mass or moment of inertia. They also establish the convergence that as the magnitude of the inertial parameter approaches infinity the natural frequencies of the constrained system are approached by the frequencies of the inertially restrained systems. As with the stiffness type penalty method, the direction of approach depends on the sign of the inertial parameter. Therefore it is possible to delimit the natural frequencies of the constrained system using asymptotic models with positive and negative inertial restraints. There is of course the choice of using positive stiffness and positive inertia penalty terms to measure and control the error due to any constraint violation but this does involve a little more work as both stiffness and mass matrices need to have penalty additions.

Let us revisit the propped cantilever, this time with an asymptotic model based on inertial penalty parameter. In simple terms this means attaching an artificial mass m_p to the free end as a penalty mass. The first three natural frequencies of a cantilever beam of length L, flexural rigidity EI and mass per unit length m, carrying a particle of mass m_p at its tip (see Figure 5.6) were determined by using the Rayleigh–Ritz method. Once again the equations in section 3.2 can be readily used to generate the results.

The variation of the non-dimensional natural frequency parameter given by $\lambda = \left(m\omega^2 L^4 / (EI)\right)^{1/4}$ against a dimensionless mass parameter $\mu = m_0 / (mL)$ is shown in Figure 5.7. The natural frequency parameter for a fully constrained propped cantilever is also shown in the same figure. The numerical results are given in Table 5.3. Figures 5.8–5.10 show how the modes for increasing inertial penalty parameters satisfy the constraints. The first mode continues to violate the constraint condition and this is the one whose natural frequency monotonically decreases. The second mode takes the shape of the first mode of a propped cantilever beam.

Comparing the results based on inertial penalty parameter with the ones from the stiffness penalty method two interesting points emerge. Firstly the magnitude of inertial penalty parameter needed to reach convergence is significantly lower than the stiffness penalty parameter. Even with a mass of about 10 times that of the beam it is possible to get accurate results, whereas with stiffness the non-dimensional parameter required would be of the order of 10,000 or greater.

Secondly, with stiffness penalty, the fundamental frequency converges first but for higher modes constraint violation persists and required even higher stiffness. For the inertial penalty parameter, the higher modes converge (in terms of satisfying constraints) before the lower ones. In fact the lowest constrained mode would be the last to converge. There is a reason for this. The inertial penalty parameter may be equivalent to a stiffness penalty parameter through the following relationship:

$$k = -\omega^2 m_p$$

Thus the penalty mass results in an equivalent stiffness penalty parameter of magnitude that is amplified by the square of the frequency. This means for higher modes, even a small inertial penalty parameter produces a very high equivalent stiffness penalty parameter!

Another interesting point to note here is that unlike the penalty stiffness, the use of penalty mass does not cause any frequencies to "go through the roof". However, one or more lower frequencies continue to decrease but will never become zero.

Numerical tests on continuous systems with rotational constraints confirm that these too can be asymptotically modeled using positive and negative artificial moments of inertia. The results are not reported here but the interactive program asymptotic.exe may be used to generate these results [ILA 00a].

The inertial penalty method is applicable for vibratory systems possessing inertia. However, any eigenvalue problem may be solved in a similar way by including a penalty term in the matrix that is associated with the eigenvalue. For example, attaching a penalty parameter to the geometric stiffness matrix in a structural stability problem is mathematically identical to applying an inertial penalty term in a vibration problem. We then need a more generic name instead of "inertial penalty" A name proposed by Ilanko and Williams is "eigenpenalty" [ILA 08].The general proof to support the use of an eigenpenalty method is given in [ILA 08].

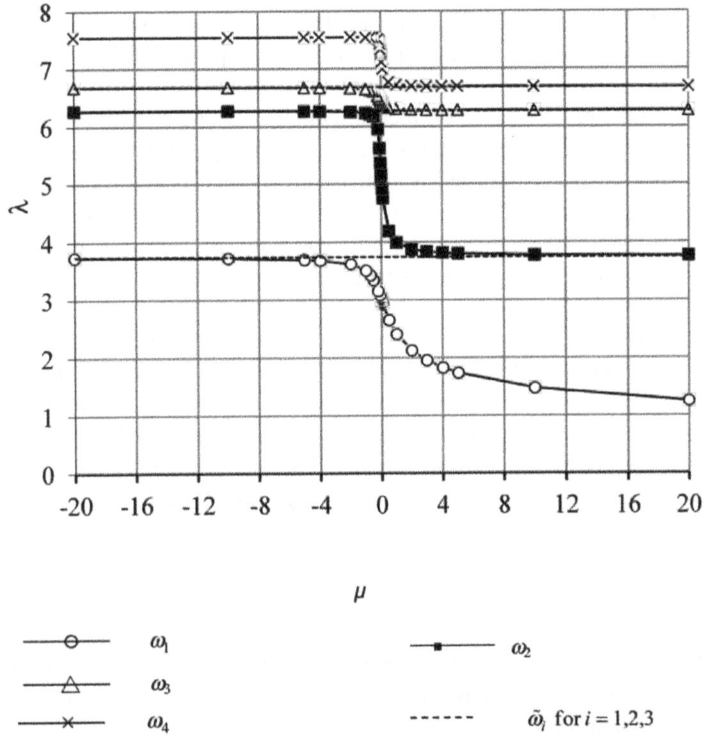

Figure 5.7. *First three frequencies of the propped cantilever and four frequencies of its asymptotic model with penalty mass*

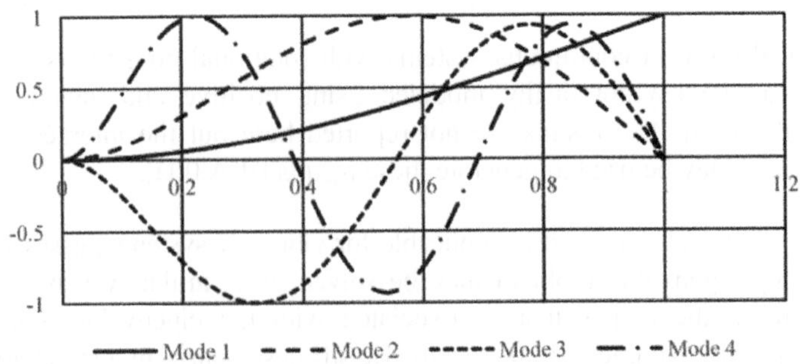

Figure 5.8. *Vibration modes for μ=0.2*

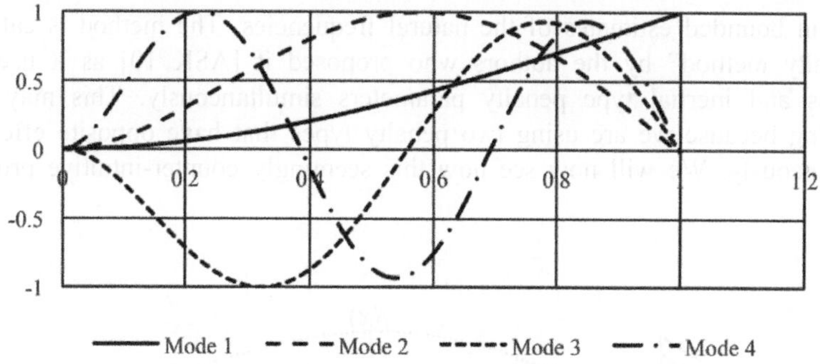

Figure 5.9. *Vibration modes for μ=1*

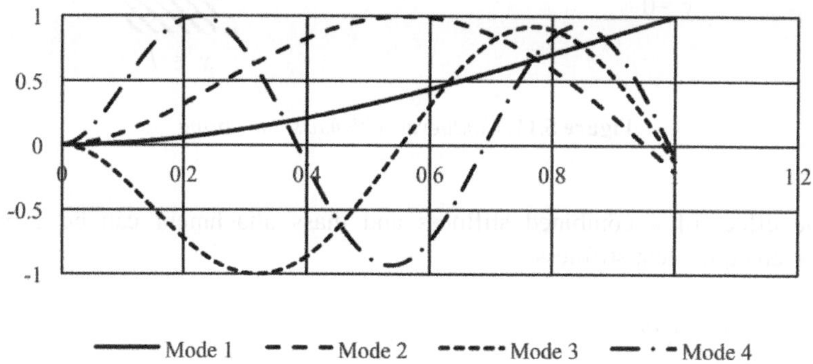

Figure 5.10. *Vibration modes for μ=10*

5.5. The bipenalty method

So we have seen that it is possible to estimate and control the error due to the approximation of the constraints with either springs of very high positive and negative stiffness, or alternatively with positive and negative mass (or inertia) of very large magnitudes. Noting that the frequencies of the constrained system are approached from below when using positive stiffness and from above when using positive mass, we can also obtain bounded results by obtaining the asymptotic solutions twice, first with positive stiffness and then with positive mass (or inertia). This is likely to be more palatable for those who do not like using negative parameters which do not have a physical representation. Interestingly, it is also possible to use a combination of stiffness and mass simultaneously (see Figure 5.11) to obtain bounded estimates of the natural frequencies. The method is called the "bipenalty method" by the authors who proposed it [ASK 10] as it uses both stiffness and inertial type penalty parameters simultaneously. This may sound confusing because we are using two penalty types that have opposite effects but simultaneously. We will now see how this seemingly counter-intuitive procedure works.

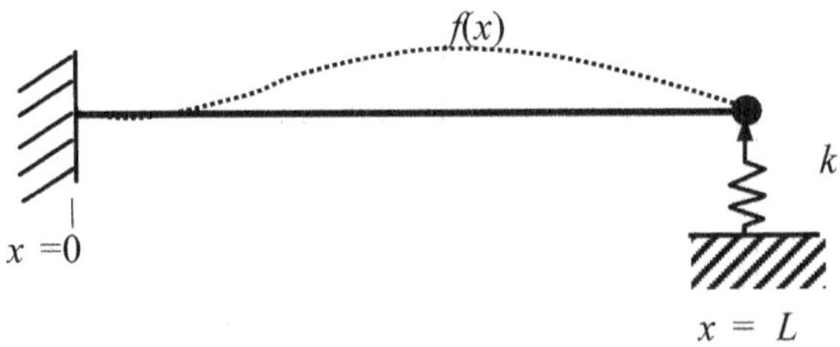

Figure 5.11. *Cantilever with mass and spring*

The effect of a combined stiffness and mass attachment can be expressed through an equivalent stiffness

$$k_e = k - m_p \omega^2 \qquad [5.1]$$

The stiffness effect would dominate if, in the range of frequency that is of interest,

$$k > m_p \omega^2 \qquad [5.2]$$

The mass effect would dominate if, in the range of frequency that is of interest:

$$k < m_p \omega^2 \qquad [5.3]$$

To ensure these inequalities, we need to know the range of interest of the frequencies.

From Rayleigh's theorem of separation, the natural frequencies of the constrained system separate the natural frequencies of the unconstrained system. As our interest is in finding the frequencies of the constrained system, we can then be sure that in the range of interest,

$$\omega \geq \omega_1 \qquad [5.4]$$

Also using Rayleigh's theorem of separation, the frequencies in the range of interest would be lower than or equal to the highest natural frequency of the constrained system. Thus,

$$\omega \leq \omega_n \qquad [5.5]$$

So the range of interest lies between the lowest and highest natural frequencies of the unconstrained system.

This means if we tune the penalty stiffness and mass to values k_1 and m_{p1} such that:

$$k_1 = m_{p1}\omega_1^2 \qquad [5.6]$$

then for any frequency in the range of interest, from equations [5.1] and [5.4], the inequality condition in equation [5.3] would be satisfied and the natural frequencies of the penalized model will give lower bounds of the natural frequencies of the corresponding constrained system.

Similarly we can see that to get upper bounds of the natural frequencies of the constrained system, the penalty parameters k_2 and m_{p2} could be tuned such that:

$$k_2 = m_{p2}\omega_n^2 \qquad [5.7]$$

With the two sets of bipenalty parameters, it is possible to obtain bounded results for the constrained system using the penalty model. The advantage of using the bipenalty method is that the very high natural frequencies corresponding to modes that violate constraints can be kept to a minimum and similarly the very small values associated with inertially penalized systems can be kept at the lowest natural frequency of the unconstrained system. Thus the frequencies will neither "go through the roof" nor "fall through the floor". These advantages are explained in [ASK 10, HET 13, ILA 12].

6

Some Useful Mathematical Derivations and Applications

6.1. Derivation of stiffness and mass matrix terms

In section 3.2 we observed that it is convenient to differentiate the energy functional with respect to the displacement coefficients G_i before carrying out any integration over the domain of the system to set up the stiffness and mass matrices. We will now derive some mathematical formulae that could be used to simplify the formation of the stiffness and mass matrices for many common one dimensional problems. Up to this point we have used φ for the exact modes of vibration and $\overline{\varphi}$ for assumed admissible functions. From here onwards, only φ will be used as we will not be concerned with the exact modes.

$$C_{i,j} = \frac{\partial^2}{\partial G_i \partial G_j} \int_{x=0}^{L} a \left(\frac{d^r f}{dx^r} \right)^2 dx \qquad [6.1]$$

where $f = \sum_{j=1,2..} G_j \varphi_j$ [6.2]

Let $z = \frac{d^r f}{dx^r} = \sum_{j=1,2..} G_j \frac{d^r \varphi_j}{dx^r}$ [6.3]

Substituting equation [6.3] into [6.1] gives:

$$C_{i,j} = \frac{\partial^2}{\partial G_i \partial G_j} \int_{x=0}^{L} a(z)^2 dx$$

$$= \frac{\partial}{\partial G_i}\int_{x=0}^{L} a\frac{\partial}{\partial G_j}(z)^2 dx = \frac{\partial}{\partial G_i}\int_{x=0}^{L} a(2z)\frac{\partial z}{\partial G_j}dx \qquad [6.4]$$

But since, $z = \sum_{j=1,2..} G_j \frac{d^r\varphi_j}{dx^r} = G_1\frac{d^r\varphi_1}{dx^r} + G_2\frac{d^r\varphi_2}{dx^r} + ... + G_j\frac{d^r\varphi_j}{dx^r} + ..$

and (G_j)s are independent,

$$\frac{\partial z}{\partial G_j} = \frac{d^r\varphi_j}{dx^r} \qquad [6.5]$$

Substituting this into equation [6.4] gives: $C_{i,j} = \frac{\partial}{\partial G_i}\int_{x=0}^{L} a(2z)\frac{d^r\varphi_j}{dx^r}dx$

Of the terms inside the integral, z is the only function that depends on G_i.

Therefore, $C_{i,j} = \int_{x=0}^{L} 2a\frac{\partial z}{\partial G_i}\frac{d^r\varphi_i}{dx^r}dx$ [6.6]

Similar to equation [6.5] we can show that, $\frac{\partial z}{\partial G_i} = \frac{d^r\varphi_i}{dx^r}$ [6.7]

Substituting this into equation [6.6] gives:

$$C_{i,j} = \int_{x=0}^{L} 2a\frac{d^r\varphi_i}{dx^r}\frac{d^r\varphi_j}{dx^r}dx \qquad [6.8]$$

This may be used to deduce simple expressions for stiffness and mass matrix coefficients. For example, the bending stiffness term may be obtained by putting r = 2, and $a = EI/2$ into equation [6.8] giving:

$$K_{i,j} = \int_{x=0}^{L} EI\frac{d^2\varphi_i}{dx^2}\frac{d^2\varphi_j}{dx^2}dx \qquad [6.9]$$

Similarly, the mass matrix coefficient due to the self-mass of a beam would be obtained by putting $r = 0$ and a = $m/2$ into equation [6.8] giving:

$$M_{i,j} = \int_{x=0}^{L} m\varphi_i\varphi_j dx \qquad [6.10]$$

Similar equations can be derived for two and three-dimensional problems but involving double or triple integrals over the domain of interest.

For concentrated masses and springs, the same procedure may be used, except in such cases one would use functions evaluated at particular co-ordinates, instead of integrals.

A mass m_0 attached at $x = a$, would result in an additional term in the mass matrix given by

$$m_0\varphi_i(a)\varphi_j(a) \qquad [6.11]$$

A spring of stiffness k that restrains the deflection $f(x)$ at $x = a$ would result in an extra term in the stiffness matrix given by

$$k\varphi_i(a)\varphi_j(a) \qquad [6.12]$$

Formulae for some common stiffness and mass matrix terms are given in the next section. The formulas for the potential energy terms associated with different types of structural elements are available in standard Mechanics of Materials text books. The derivation for the potential energy associated with the axial force which gives rise to what is known as the geometric stiffness is not available in most text books at present. The derivation available appears to be based on the assumption that one end of the beam is free to move axially [MCC 73] although the final result is applicable for any axial boundary condition for small amplitude vibrations. The derivation for this potential energy term is given in Chapter 13.

6.2. Frequently used potential and kinetic energy terms

Table 6.1 gives the potential and kinetic energy terms for some common vibratory systems. The following notation is used:

E - Elastic modulus G - Shear modulus ν - Poisson's ratio

A - Cross sectional area I - Second moment of area @ the neutral axis

J - Polar second moment of area of a circular cross section

P - Static axial force in a beam g - gravitational acceleration

k - Stiffness of a spring k_f - Stiffness of an elastic foundation / unit length

m_0 - Magnitude of a concentrated mass

I_0 - Moment of inertia of a mass about the relevant axis.

u - Axial displacement | x - Axial co-ordinate

w- Lateral displacement | θ- Angle of twist

w', u' - First derivatives of w,u | w'' - Second derivative of w

ω- Frequency of vibration. | e - elongation /shortening of a spring

NOTE.– In the application of Rayleigh's principle, or the Rayleigh–Ritz method (RRM), the terms u,w and θ should be taken as the displacement/rotation at the time of maximum excursion. Maximum translational velocities ($\partial v/\partial t$), ($\partial u/\partial t$) etc. are given by ωf and maximum flexural rotational velocity ($\partial w'/\partial t$) is given by $\omega f'$.

Case type	Potential energy	Stiffness matrix ($K_{i,j}$)
Axial strain energy in a bar	$(\frac{1}{2})\int EA(u')^2\,dx$	$\int_0^L EA\varphi_i'\varphi_j'dx$
Torsional strain energy of a circular shaft	$(\frac{1}{2})\int GJ(\theta')^2\,dx$	$\int_0^L GJ\varphi_i'\varphi_j'dx$
Strain energy due to flexure	$(\frac{1}{2})\int EI(w'')^2\,dx$	$\int_0^L EI\varphi_i''\varphi_j''dx$
Additional potential energy due to axial force in beams	$(\frac{1}{2})\int P(w')^2\,dx$	$\int_0^L P\varphi_i'\varphi_j'dx$
Strain energy in a spring	$(\frac{1}{2})\,ke^2$	$k\varphi_i(a)\varphi_j(a)$ if $e = v\,(a)$
Strain energy in an elastic foundation	$(\frac{1}{2})\int k_f w^2\,dx$	$\int_0^L k_f\varphi_i\varphi_j dx$
Case type	**Kinetic energy**	**Mass matrix ($M_{i,j}$)**
Longitudinal vibration of a bar	$(\frac{1}{2})\int \rho A(\partial u/\partial t)^2\,dx$	$\int_0^L \rho A\varphi_i\varphi_j dx$
Lateral vibration of a beam	$(\frac{1}{2})\int \rho A(\partial w/\partial t)^2\,dx$	
Torsional vibration of a shaft	$(\frac{1}{2})\int \rho J(\partial\theta/\partial t)^2\,dx$	$\int_0^L \rho J\varphi_i\varphi_j dx$
Rotary inertia of a beam	$(\frac{1}{2})\int \rho I(\partial w'/\partial t)^2\,dx$	$\int_0^L \rho I\varphi_i'\varphi_j'dx$
Translation of concentrated mass m_0 at $x = a$	$(\frac{1}{2})\,m_0(\partial w\partial t)^2$	$m_0\varphi_i(a)\varphi_j(a)$
Rotation of a thin rigid body (w' or θ) with inertia I_0 at $x = a$	$(\frac{1}{2})\ I_0\ (\partial w'/\partial t)^2$ or $(\frac{1}{2})\ I_0(\partial\theta/\partial t)^2$	$I_0\varphi_i'(a)\varphi_j'(a)$

Table 6.1. *Stiffness & mass matrix elements for some common energy terms*

6.3. Rigid body connected to a beam

The formulas given in the preceding table should be used with caution. If the thickness of an added mass is not negligible then the location of its centroid will affect the kinetic energy. To illustrate this let us consider the rigid connection of a rigid body having a mass m_0 and moment of inertia I_0 (about its relevant centroidal axis) to the end of cantilever beam.

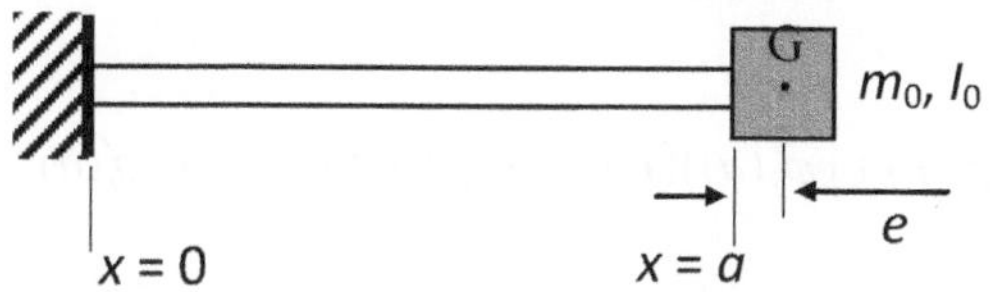

Figure 6.1. *Cantilever carrying a rigid end mass*

Let the distance between its centroid and the beam be e.

As usual let $f(x) = \sum_j G_j \varphi_j(x)$

Then $f(a) = \sum_j G_j \varphi_j(a)$, and $f'(a) = \sum_j G_j \varphi'_j(a)$

For small rotations, the translation of the centre of the body is given by $f(a) + e$ $f'(a)$.

The kinetic energy function due to the vibration of the rigid body is:

$$\psi_m = \frac{m_0}{2}\left(f(a) + ef'(a)\right)^2 + \frac{I_0}{2}\left(f'(a)\right)^2$$

$$\frac{\partial \psi_m}{\partial G_j} = m_0\left(f(a) + ef'(a)\right)\frac{\partial}{\partial G_j}\left(f(a) + ef'(a)\right) + I_0 f'(a)\frac{\partial}{\partial G_j} f'(a)$$

Substituting $\frac{\partial f(a)}{\partial G_j} = \varphi_j(a)$ and $\frac{\partial f'(a)}{\partial G_j} = \varphi'_j(a)$ into the above we get,

$$\frac{\partial \psi_m}{\partial G_j} = m_0 \left(f(a) + ef'(a) \right) \left(\varphi_j(a) + e\varphi_j'(a) \right) + I_0 f'(a) \varphi_j'(a)$$

$$\frac{\partial^2 \psi_m}{\partial G_i \partial G_j} = m_0 \left(\varphi_j(a) + e\varphi_j'(a) \right) \frac{\partial}{\partial G_i} \left(f(a) + ef'(a) \right) + I_0 \varphi_j'(a) \frac{\partial f'(a)}{\partial G_i}$$

Since $\dfrac{\partial f(a)}{\partial G_i} = \varphi_i(a)$ and $\dfrac{\partial f'(a)}{\partial G_i} = \varphi_i'(a)$ the above equation reduces to:

$$\frac{\partial^2 \psi_m}{\partial G_i \partial G_j} = m_0 \left(\varphi_j(a) + e\varphi_j'(a) \right) \left(\varphi_i(a) + e\varphi_i'(a) \right) + I_0 \varphi_j'(a) \varphi_i'(a)$$

Therefore the additional term in the mass matrix is:

$$m_0 \left(\varphi_j(a) + e\varphi_j'(a) \right) \left(\varphi_i(a) + e\varphi_i'(a) \right) + I_0 \varphi_j'(a) \varphi_i'(a) \quad [6.13]$$

If the mass were connected to the beam at a midpoint then it is advisable to split the beam into two segments and enforce geometric continuity at the connections either using the Lagrangian multiplier method, or by inserting artificial springs with high stiffness to effectively constrain any relative translation or rotation at the connections.

6.4. Finding the critical loads of a beam

Having discussed examples in vibration, let us see how the RRM could be used to find the static critical loads of beams subject to constraints. Once again, let us start with a cantilever and then impose geometric constraints against translation at the free end using a penalty term. Thus the task is to find the values of P at which the beam is in critical equilibrium.

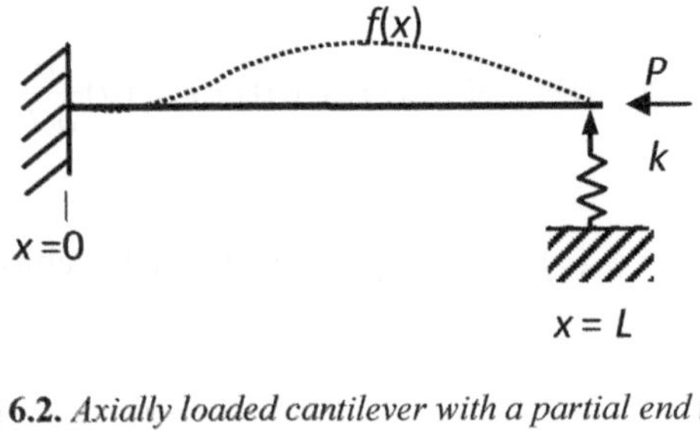

Figure 6.2. *Axially loaded cantilever with a partial end restraint*

There are only stiffness matrices, one associated with elastic bending stiffness which is identical to the vibration case. The geometric stiffness matrix which is associated with the axial load *P* as the eigenvalue can be obtained from Table 6.1.

The total potential energy is given by:

$$V = \int_0^L \frac{EI}{2}\left(\frac{d^2w}{dx^2}\right)^2 dx - \int_0^L \frac{P}{2}\left(\frac{dw}{dx}\right)^2 dx + (k/2)(w(L))^2 \qquad [6.14]$$

Minimization leads to three matrices. The elastic stiffness matrix **K**, the geometric stiffness matrix $\mathbf{K}^*$ and the penalty stiffness matrix $\mathbf{K_p}$.

From Table 6.1,

$$K_{i,j} = \int_0^L EI\varphi_i''\varphi_j''dx \qquad [6.15]$$

$$K_{i,j}' = -\int_0^L P\varphi_i'\varphi_2'dx \qquad [6.16]$$

In this case, *P* is an unknown and it is therefore convenient to express this in terms of a load independent "geometric stiffness matrix" **K*** as follows:

$$K_{i,j}' == -PK_{i,j}^* = -P\int_0^L \varphi_i'\varphi_2'dx \text{, where} \qquad [6.17a]$$

$$K_{i,j}^* = \int_0^L \varphi_i'\varphi_2'dx \qquad [6.17b]$$

Using the same admissible functions that were used for the dynamic displacement in section 3.2,

$$w(x) = \sum\nolimits_{j=1,2}^{n} G_j\bar{\varphi}_j \text{ where } \bar{\varphi}_j = (x/L)^{j+1} \qquad [6.18]$$

$$K_{i,j} = \int_0^L EI\varphi_i''\varphi_j''dx =$$

$$\int_0^L EI(i+1)i(j+1)j\left(\frac{x^{i+j-2}}{L^{i+j+2}}\right)dx = \frac{(i+1)i(j+1)jEI}{(i+j-1)L^3} \qquad [6.19]$$

$$K_{i,j}^{*} = \int_0^L \varphi_i' \varphi_2' dx$$

$$= \int_0^L (i+1)(j+1)\left(\frac{x^{i+j}}{L^{i+j+2}}\right)dx = \frac{(i+1)(j+1)}{(i+j+1)L} \qquad [6.20]$$

$$K_{p,i,j} = k\varphi_i(L)\varphi_j(L) = k \qquad [6.21]$$

Then the eigenvalue equation is of the form:

$$[\mathbf{K}+\mathbf{K}_\mathbf{p}]\{\mathbf{G}\} - P[\mathbf{K}^{*}]\{\mathbf{G}\} = \{0\} \qquad [6.22]$$

Single term solution:

Taking only one term (n=1 in equation [6.18]) we have:

$$[K_{1,1} + K_{p,1,1}]\{G_1\} - P[K_{1,1}^{*}]\{G_1\} = \{0\} \qquad [6.23]$$

$$K_{1,1} = 4EI / L^3; K_{p,1,1} = k; K_{1,1}^{*} = 4/(3L)$$

Substituting into equation [6.23] gives:

$$\left(4EI / L^3 + k - 4P/(3L)\right)G_1 = 0$$

There are two solutions.

Either $G_1 = 0$, (this means there is no displacement and the beam is in equilibrium)

or $\left(4EI / L^3 + k - 4P/(3L)\right) = 0$

The second case means G_1 is non-zero and may take any value. Thus a critical state occurs, because the displacement is indefinite as G_1 can take any value and the beam would be in equilibrium. The critical load is thus:

$$P = 3EI / L^2 + 3kL / 4$$

If $k = 0$, then the critical load of the cantilever is given by $P = 3EI / L^2$.

The exact value of the lowest critical load of a cantilever is given by $P = \pi^2 EI / (4L^2)$.

The error is 21.5%.

The mode is given by the first term in equation [6.18] which is a square term. The critical load would also increase monotonically with the stiffness k and therefore the single term solution will not give the critical load of a propped cantilever. The results obtained using a MATLAB program (see appendix A.3) for multiterm solutions are given in Table 6.2 for $k = 0$ (cantilever). Some of the modes could not be found (these are marked with a-) because of inadequate number of terms. For example with $n = 1$, it is possible to find only one critical load. Figure 6.3 shows the buckling modes.

	Mode 1	Mode 2	Mode 3	Mode 4
1	3	–	–	–
2	2.486	32.181	–	–
3	2.468	23.391	109.141	–
4	2.467	22.322	69.404	265.807
5	2.467	22.214	63.028	148.205
6	2.467	22.207	61.863	127.207
7	2.467	22.207	61.702	122.167
8	2.467	22.207	61.686	121.097
9	2.467	22.207	61.685	120.925
Exact	2.467	22.207	61.685	120.903

Table 6.2. *Critical loads of a cantilever*

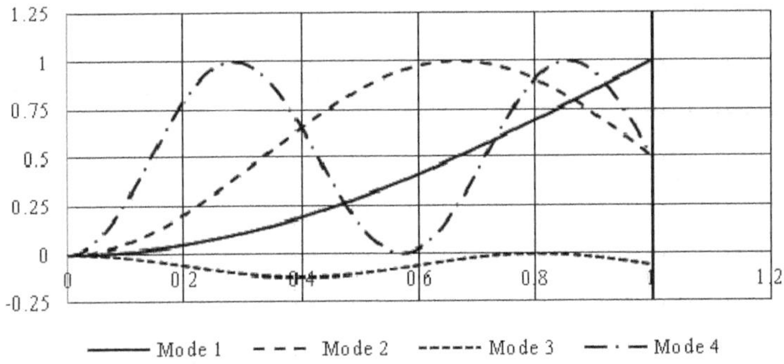

Figure 6.3. *Buckling modes of a cantilever*

Table 6.3 gives the critical loads for k=$10^6EI/L^3$. This corresponds to a propped cantilever. Note in Table 6.3, the large penalty term leads to a very large value for the highest critical load which corresponds to a constraint violation. These are marked with * in the Table. The buckling modes are shown in Figure 6.4. Please note that the penalty functions have been effective in enforcing zero displacement at the right end.

	Mode 1	Mode 2	Mode 3	Mode 4
1	*	-	-	-
2	30.000	*	-	-
3	20.919	107.081	*	-
4	20.286	65.931	263.783	*
5	20.194	60.972	143.091	543.741
6	20.191	59.813	125.146	264.204
7	20.191	59.694	119.977	217.387
8	20.191	59.680	119.087	202.456
9	20.191	59.679	118.920	198.928
Exact	20.191	59.680	118.900	197.858

Table 6.3. *Critical Loads of a propped cantilever beam*

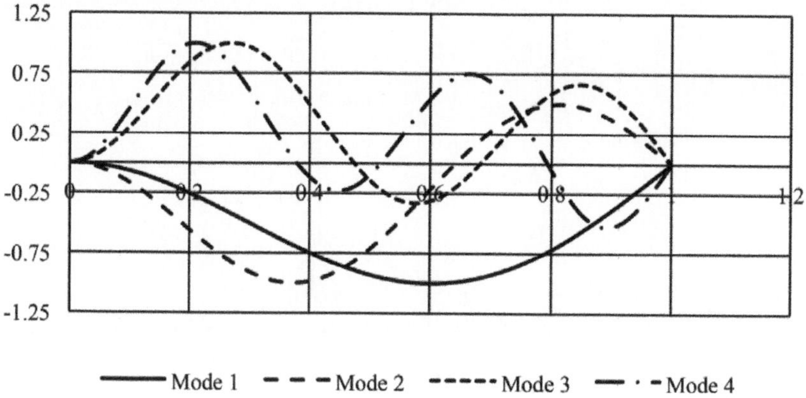

Figure 6.4. *Buckling modes of a propped cantilever*

7

The Theorem of Separation and Asymptotic Modeling Theorems

7.1. Rayleigh's theorem of separation and the basis of the Ritz method

Until now we have been focusing on how to apply the RRM and the penalty method. In this section we will attempt to present the theoretical basis of these methods. In section 2.2, it was shown that the application of the principle of conservation of energy gives an upper bound estimate of the fundamental natural frequency. Based on this we sought to get the best possible estimate of the fundamental natural frequency by minimizing the Rayleigh quotient resulting in the Rayleigh–Ritz equations. We have seen that this procedure gives estimates of more than one natural frequency, the number being equal to the number of terms used. In a cantilever beam example, we have also observed that the higher frequencies calculated were never lower than the exact frequencies, and converged toward the exact solution with number of terms used. There is another good piece of news here. This is not just an observation. There are theorems to show that the Rayleigh–Ritz minimization gives upper bound estimates for the higher natural frequencies too! These are described below. The proof presented here is taken from an excellent book on variational principles by Gould [GOU 66] but the notation used has been changed to make it more user-friendly for engineers, and the intermediate steps and explanations have been revised.

THEOREM (a).– The lowest natural frequency of a system cannot decrease as a result of adding one or more constraints; if the number of additional constraints is h, the lowest natural frequency cannot exceed the $(h+1)$th natural frequency of the original system.

i.e. $\omega_1 \le \tilde{\omega}_{1,h} \le \omega_{h+1}$

where ω_1 and ω_{h+1} are the natural frequencies of the original system (A) corresponding to the fundamental mode and the $(h+1)^{\text{th}}$ mode respectively and $\tilde{\omega}_{1,h}$ is the lowest natural frequency of the constrained system $\tilde{A}_h$.

PROOF.–

By definition, $\omega_1^2 = \min\left(\dfrac{V_m}{\psi_m}\right)$ [7.1]

where V_m and ψ_m are the maximum total potential energy and maximum kinetic energy function of the original system.

$$\tilde{\omega}_{1,h}^2 = \min\left(\frac{V_m}{\psi_m}\right)$$

subject to constraint conditions

$$\bar{f}_i = o \text{ for } i=1,2,3..h \qquad [7.2]$$

Compared to equation [7.1], $\tilde{\omega}_{1,h}$ is obtained using a (more) restricted choice of displacement functions u. Since the minimum from a more restricted choice of u cannot be less than the minimum from a less restricted choice (principle of monotony),

$$\omega_1^2 \le \tilde{\omega}_{1,h}^2 \qquad [7.3]$$

Let us now introduce a parameter Ω_1 which may be defined as an estimate of the fundamental natural frequency of the constrained system obtained by taking the minimum of (V_m/ψ_m), using a linear combination of the first $(h+1)$ modes of the original system, but the sum of the series subject to h constraints.

$$\Omega_1^2 = \frac{\sum_{i=1}^{h+1} a_i^2 \omega_i^2 M_i}{\sum_{i=1}^{h+1} a_i^2 M_i}$$

Dividing both sides by ω_{h+1}^2 gives:

$$\left(\frac{\Omega_1}{\omega_{h+1}}\right)^2 = \frac{\sum_{i=1}^{h+1} a_i^2 (\omega_i/\omega_{h+1})^2 M_i}{\sum_{i=1}^{h+1} a_i^2 M_i} \le 1$$

Therefore $\Omega_1^2 \leq \omega_{h+1}^2$ [7.4]

Note that $(h+1)$ is the minimum number of modes required to obtain a displacement form that satisfies h constraints. $\tilde{\omega}_{1,h}^2$ is the minimum of (V_m / ψ_m) from a wider selection of u compared to that used in finding Ω_1^2.

Therefore, $\tilde{\omega}_{1,h}^2 \leq \Omega_1^2$ [7.5]

From inequalities [7.3] and [7.5] we get $\omega_1 \leq \tilde{\omega}_{1,h} \leq \Omega_1 \leq \omega_{h+1}$ which gives:

$$\omega_1 \leq \tilde{\omega}_{1,h} \leq \omega_{h+1} \qquad [7.6]$$

This result has an interesting implication. The maximum possible value for the fundamental natural frequency with h constraints is the $(h+1)$th natural frequency of the original system.

That is to say $(\tilde{\omega}_{1,h})_{\max} = \omega_{h+1}$ [7.6a]

This means the constraints that give the greatest increase in the lowest natural frequency are the h orthogonality conditions relating to the first h modes of the system. It may also be said that, where nodes can be found, the greatest gain in the fundamental natural frequency of a system may be achieved by placing the h constraints on the nodes of the $(h+1)$th mode.

$(\tilde{\omega}_{1,h})_{\max} = \min(V_m / \psi_m)$ subject to conditions

$$\overline{m}u \bullet u_i = 0 \text{ for } i = 1, 2 \ldots h \qquad [7.6b]$$

where u_i is the ith mode of the original system.

THEOREM (b).– The rth natural frequency of the system with h constraints is bounded by the rth natural frequency and the rth natural frequency of the original system. If $\omega_r, \tilde{\omega}_{r,h}$, and ω_{r+h} are the natural frequencies corresponding to the rth mode of the original system, rth mode of the system with h constraints and $(r+h)$th mode of the original system, respectively, then $\omega_r \leq \tilde{\omega}_{r,h} \leq \omega_{r+h}$

PROOF.–

$$\omega_r^2 = \min(V_m / \psi_m)$$

subject to orthogonality conditions:

$$\bar{m}u \bullet u_i = 0 \text{ for } i = 1, 2...(r-1) \quad [7.7]$$

Let $\Omega_r^2 = \min(V_m / \psi_m)$

subject to orthogonality conditions: $\bar{m}u \bullet u_i = 0$ for $i = 1, 2...(r-1)$ and constraint conditions:

$$f_i=0, \text{ for } i = 1,2,\ldots h \quad [7.8]$$

Conditions under which the minimum is obtained in [7.8] are stronger than that in [7.7].

$$\text{Therefore, } \omega_r^2 \leq \Omega_r^2 \quad [7.9]$$

If we now ask the question, which r constraint conditions would result in the greatest gain in the Quotient Ω_r^2 for the system $\tilde{A}_h$, from the implications of Theorem (a) as explained already, the answer would be the r orthogonality conditions for $\tilde{A}_h$. That is $\bar{m}u \bullet \tilde{u}_i = 0$ for $i = 1, 2...(r-1)$ would result in the maximum possible increase for the fundamental frequency of the constrained system.

If $\tilde{\omega}_{r,h}^2 = \min(V_m / \psi_m)$

subject to orthogonality conditions: $\bar{m}u \bullet \tilde{u}_i = 0$ for $i = 1, 2...(r-1)$

and constraint conditions:

$$f_i=0, \text{ for } i = 1,2,\ldots h \quad [7.10]$$

then $\Omega_r^2 \leq \tilde{\omega}_{r,h}^2$ [7.11]

Using this and the inequality (9) gives:

$$\omega_r^2 \leq \tilde{\omega}_{r,h}^2 \quad [7.12]$$

This means the *r*th natural frequency of a system cannot be decreased by adding one or more constraints to that system. Let us now find an estimate of the *r*th natural frequency, say $\tilde{\Omega}_{r,h}$, of a system subject to h constraints ($\tilde{A}_h$) by using a linear combination of the first (r+h) modes of the original system (A) for the displacement form.

i.e. Let $u = \sum_{i=1}^{r+h} a_i u_i$

It may be noted that $(r+h)$ terms are sufficient to ensure h constraints and $(r\text{-}1)$ orthogonality conditions.

$$\tilde{\Omega}_{r,h}^2 = \min(V_m / \psi_m) = \frac{\sum_{i=1}^{r+h} a_i^2 \omega_i^2 M_i}{\sum_{i=1}^{r+h} a_i^2 M_i} \leq \omega_{r+h}^2 \qquad [7.13]$$

If we do not restrict the choice of shapes to the $(r+h)$ modes of (A), we will get a value which cannot be greater than $\tilde{\Omega}_{r,h}$

$$\text{i.e. } \tilde{\omega}_{r,h}^2 \leq \tilde{\Omega}_{r,h}^2 \qquad [7.14]$$

Combining inequalities [7.12]–[7.14] gives:

$$\omega_r \leq \tilde{\omega}_{r,h} \leq \omega_{r+h} \qquad [7.15]$$

This theorem forms the basis of the RRM where a series of admissible displacement forms are used to represent the vibrational displacement of a system. Any chosen series may be considered as a constrained form of the actual displacement shape and therefore, from the left-hand-side inequality in equation [7.15], we may conclude that the Rayleigh–Ritz minimization leads to upper bound estimates of all the natural frequencies except in the case of a highest mode of a discrete system.

7.2. Proof of convergence in asymptotic modeling

As mentioned earlier, the major drawback of the RRM is the difficulty in selecting admissible functions that satisfy all geometric constraints. We have seen that by using artificial springs with large stiffness we can overcome this problem. We have also seen that the use of negative and positive stiffness enables us to control any error due to the asymptotic modeling itself. When the author first suggested the use of negative stiffness, questions were raised about the possibility of introducing instability and the lack of proof that the negative stiffness will give true upper bound estimates for the constrained systems. Further work using Rayleigh's theorem of separation has led to the proof of two other theorems, which justify the use of very large positive and negative stiffness to effectively impose constraint conditions. These are given in this section which is a condensed version of the author's proof published in the Journal of Sound and Vibration [ILA 02a].

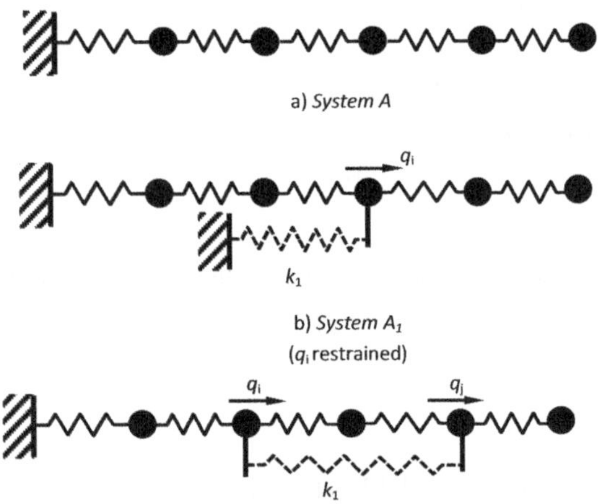

Figure 7.1. *Schematic of a discrete system and asymptotic models of the same system subjected to absolute and relative constraints. Penalty springs sketched with dotted lines*

7.2.1. *The natural frequencies of an n DOF system with one additional positive or negative restraint*

Consider an n degree of freedom system. We will call this the "original system or System A. Figure 7.1(a) shows a typical spring-mass system. We are using this only as an illustration but the derivations hold for any type of discrete system. Let us now add to this system, an artificial spring of stiffness k_1. The expression for the total potential energy of the system will now have an additional term

$$\Delta V = \frac{1}{2} k_1 q_i^2 \qquad [7.16a]$$

where q_i is the displacement that is being restrained by (resisted by) the artificial spring. Thus in order to move by q_i the spring has to be given a force $k_1 q_i$. See Figure 7.1(b). In some situations we may want to constrain not an absolute displacement but a relative displacement. For example, if we want to restrain the relative displacement of two co-ordinates q_i and q_j, with a spring as shown in Figure 7.1(c), the change in potential energy will take the form:

$$\Delta V = \frac{1}{2} k_1 (q_i - q_j)^2 \quad [7.16b]$$

Let us refer to the new modified system, which has been obtained by modifying System A by adding one spring, as "restrained system" or System A_1.

Using positive values for the spring stiffness, since increasing the stiffness cannot decrease the natural frequencies [RAY 45] the natural frequencies of the modified system are upper bounds to the natural frequencies of the original system.

$$\omega_{m,1} \geq \omega_m \text{ for } k_1 > 0 \quad [7.17]$$

The natural frequencies may be expressed in terms of kinetic and potential energies as follows:

$$(\omega_m)^2 = \min\left(\frac{V}{\psi}\right) \text{ subject to } \bar{m}.u \bullet u_i = 0 \text{, for } i = 1,2..(m\text{-}1) \quad [7.18a]$$

$$(\omega_{m,1})^2 = \min\left(\frac{V + \Delta V}{\psi}\right) \text{ subject to } \bar{m}.u \bullet u_i = 0 \text{, for } i = 1,2..(m\text{-}1) \quad [7.18b]$$

where V is the potential energy of System A, and Ψ is a kinetic energy function for Systems A or A_1. The kinetic energy function consists of terms such as $(1/2)\bar{m}_i q_i^2$ and is of the same form for both systems, although its actual value would depend on the displacement form u that is used in the analysis. The statement $\bar{m}.u \bullet u_i = 0$, for i = 1,2..(m-1) refers to the conditions of orthogonality of the displacement form with respect to the first (m-1) modes. We could also take this to mean that the displaced form will not have any component in the first (m-1) modes.

The highest natural frequency of System A_1 ($\omega_{n,1}$) will increase monotonically with stiffness k_1 and,

$$\text{as } k_1 \to \infty, \qquad \omega_{n,1} \to \infty \quad [7.19]$$

However this does not happen with other natural frequencies for the following reasons.

As the stiffness approaches infinity, equation [7.18b] can be satisfied in two ways.

One is that the change in the potential energy ΔV, and hence the right hand side of equation [7.18b], may also approach infinity. This would make the frequency also approach infinity.

The alternative is that ΔV takes a definite value. How could this be possible? The answer is simple. If one term in a product becomes infinite the other term has to vanish. So as a result of the displacement q_i or the relative displacement $(q_i - q_j)$ becoming zero, the change in energy may be finite. In this case, finite values for natural frequencies are obtainable. This case is of course the one we are interested in because it represents a system where a displacement of the point that is attached to the spring (or the relative displacement between two points that are attached to the spring) is effectively being forced to remain zero. The stiffness of the spring acts as a penalty against any displacement (or relative displacement). It will be shown that this convergence to the constrained system occurs for (n-1) modes of a system with one added restraint.

Considering the limiting case where k_1 approaches infinity, the constrained system that is being approached will have n-1 degrees of freedom, and hence n-1 natural modes and frequencies. Let us refer to this system, which is obtained from the original system by adding a constraint corresponding to the spring restraint, as constrained system or System $\tilde{A}_1$. We may consider the asymptotic model as slightly less stiff compared to the constrained System $\tilde{A}_1$. Therefore all n–1 natural frequencies of the asymptotic model will be equal to, or slightly lower than, the corresponding frequencies of the constrained System. Denoting the mth natural frequency of System $\tilde{A}_1$ by $\tilde{\omega}_{m,1}$, we may state this as

$$\omega_{m,1} \le \tilde{\omega}_{m,1}, \quad \text{for } m < n \text{ and } k_1 > 0 \qquad [7.20]$$

Applying Rayleigh's theorem of separation to System A we get,

$$\omega_m \le \tilde{\omega}_{m,1} \le \omega_{m+1}, \quad \text{for } 1 \le m < n. \qquad [7.21]$$

This is illustrated in Figure 7.2 which is reproduced here for convenience. The natural frequencies of the original System (A) are located at the vertical axis through zero stiffness and the penalized solution represented by lines with markers can be seen to approach the natural frequencies of System $\tilde{A}_1$ from different directions depending on the sign of the stiffness. It should be noted here that while the inequality in equation [7.21] is applicable for any constraint, we are only interested in the particular case where the constraint corresponds to the restraint in System A_1.

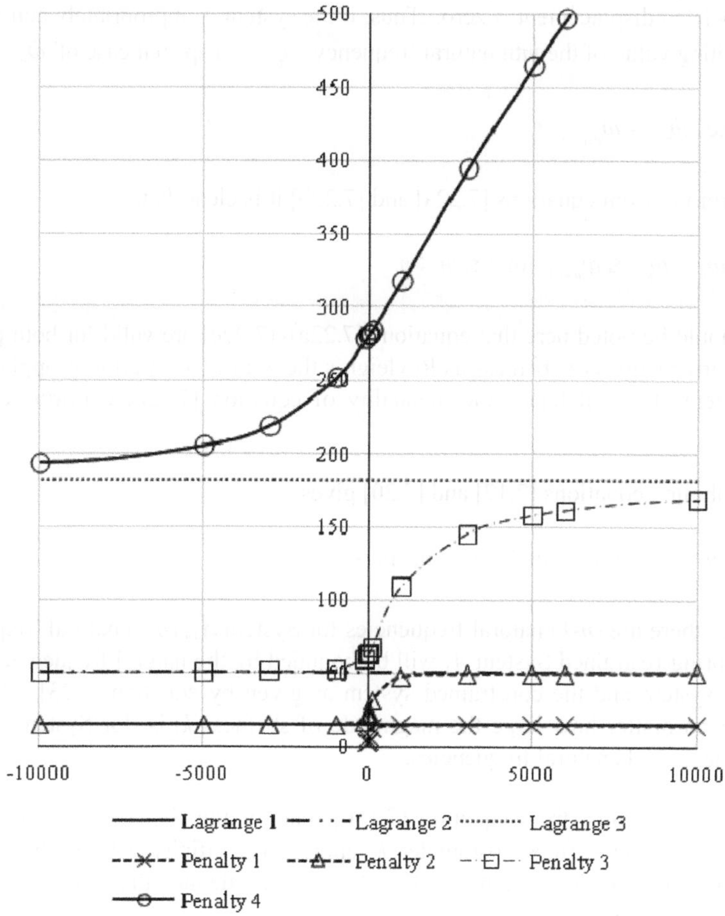

Figure 7.2. *Variation of the frequency parameter* λ^2 *with added stiffness*

We can also apply Rayleigh's theorem of separation to System A_1. This would give:

$$\omega_{m,1} \leq \tilde{\omega}^*_{m,1} \leq \omega_{m+1,1} \quad \text{for } 1 \leq m < n. \qquad [7.22a]$$

where $\tilde{\omega}^*_{m,1}$ is the mth natural frequency of System A_1 subject to *any* additional constraint. If however, the additional constraint corresponds to the restraint (spring) in A_1, then for this particular constrained system (we will refer to this as "the appropriately constrained system"), the restraint (spring) is redundant, as the associated displacement is zero. Thus, if the system is appropriately constrained, the resulting value of the mth natural frequency $\tilde{\omega}_{m,1}$ is a special case of $\tilde{\omega}^*_{m,1}$.

$$\text{i.e. } \tilde{\omega}_{m,1} \in \tilde{\omega}^*_{m,1} \qquad [7.22b]$$

Therefore, from equations [7.22a] and [7.22b] it is clear that,

$$\omega_m \leq \tilde{\omega}_{m,1} \leq \omega_{m+1,1} \quad \text{for } 1 \leq m < n. \qquad [7.22c]$$

It should be noted here that equations [7.22a]–[7.22c] are valid for both positive and negative values of stiffness as Rayleigh's theorem of separation is applicable to any system. The left hand side inequality of equation [7.22c] confirms equation [7.20].

Combining equations [7.17] and [7.20] gives

$$\omega_m \leq \omega_{m,1} \leq \tilde{\omega}_{m,1} \text{ for } 1 \leq m < n, \text{ and } k_1 > 0 \qquad [7.23]$$

Since there are (n-1) natural frequencies for System $\tilde{A}_1$, (n-1) natural frequencies of the spring-restrained System A_1 will be bounded by the natural frequencies of the original system and the constrained system as given by equation [7.23]. This is to say that no matter how large the magnitude of stiffness k_1 is, for System A_1 there exist at least (n-1) natural frequencies.

This also means, from equation [7.18b] that for the first (n-1) modes, as the magnitude of the stiffness parameter k_1 approaches infinity, $(\Delta V/\psi)$ has a limit. From this and equations [7.16a] or [7.16b], we may deduce that for the first (n-1) modes, as $k_l \rightarrow \infty$,

$$\text{either } q_i \rightarrow 0, \qquad [7.24a]$$

$$\text{or } (q_i - q_j) \rightarrow 0, \qquad [7.24b]$$

depending on whether the spring restrains a single displacement or a relative displacement between two points, as otherwise $(\Delta V/\psi) \rightarrow \infty$ which is only true for the nth mode.

This means for the first $(n\text{-}1)$ modes,

$$\text{as } k_l \rightarrow \infty, \; A_l \rightarrow \tilde{A}_1 \qquad [7.24c]$$

Using the right hand side inequality of equation [7.23] and the fact that $\omega_{m,1}$ cannot decrease with increasing k_1, we can state that,

$$\text{as } k_1 \rightarrow \infty, \qquad \omega_{m,1} \rightarrow \tilde{\omega}_{m,1}, \text{ for } 1 \leq m < n \qquad [7.25]$$

For very high values of stiffness, these natural frequencies approach that of the constrained system as given by equation [7.25]. However, the highest natural frequency of System A_1 has no such bounds, and increases monotonically with stiffness as given by equation [7.29]. A sketch of the variation of the natural frequencies of System A_1 with the stiffness parameter k_1 is shown in Figure 7.2.

Let us now consider the case of adding a restraint (for example, a spring) having a negative stiffness. If, for system A_1, the stiffness parameter k_1 were assigned a negative value, then the resulting change in the potential energy would be negative. This means that the natural frequencies of System A_1 would be less than or equal to the corresponding frequencies of the original system.

$$\text{If } \omega_{m,1} \text{ exists, } \omega_{m,1} \leq \omega_m, \text{ for } k_1 < 0 \text{ and } m \leq n \qquad [7.26]$$

Once again, the use of very large values for the magnitude of the spring stiffness can be accommodated in two different ways. As the magnitude of negative stiffness is increased, ΔV may continue to decrease, and if the total potential energy $V + \Delta V$ becomes zero, the eigenvalue would be zero, indicating a state of critical equilibrium. That is to say, the square of the lowest natural frequency of the system with a spring having a negative stiffness will monotonically decrease, becoming negative at one stage. Beyond this point the square root of the lowest eigenvalue will therefore be imaginary and the corresponding mode will cease to exist. For very high value of negative stiffness, the lowest eigenvalue will not correspond to a natural frequency. However, for System A_1, it will now be shown that this only happens to one mode, and there exist $(n\text{-}1)$ natural frequencies and modes, which asymptotically approach the natural frequencies and modes of the corresponding constrained system.

Since a decrease in stiffness cannot cause an increase in natural frequencies,

$$\text{if } \omega_{m+1,1} \text{ exists } \omega_{m+1,1} \leq \omega_{m+1}, \text{ for } 1 \leq m < n \text{ and for } k_1 < 0. \qquad [7.27a]$$

Combining equation [7.27a] and the right hand side inequality in equation [7.22c] gives:

$$\tilde{\omega}_{m,1} \leq \omega_{m+1,1} \leq \omega_{m+1}\text{, for } 1 \leq m < n \text{ and } k_1 < 0. \qquad [7.27b]$$

From the above equation it is clear that since there are (n-1) natural frequencies ($\tilde{\omega}_{m,1}$) for the constrained system, there will also exist at least (n-1) natural frequencies for the system with one negative spring. Similar to the arguments used for the positive stiffness case, for at least (n-1) modes,

as $k_l \rightarrow -\infty$, either $q_i \rightarrow 0$, [7.28a]

or $(q_i - q_j) \rightarrow 0$, [7.28b]

This means for (n-1) modes, as $k_l \rightarrow -\infty$, $A_l \rightarrow \tilde{A}_1$ [7.28c]

Since the natural frequencies cannot increase with any decrease in stiffness, using equations [7.27b] and [7.28c]

$$\text{as } k_1 \rightarrow -\infty, \qquad \omega_{m+1,1} \rightarrow \tilde{\omega}_{m,1} \text{ for } 1 \leq m < n \qquad [7.29]$$

Note that the lowest natural frequency of the constrained system occurs when m =1, and this is approached by the second mode of the system with spring having negative stiffness. The first natural frequency of the spring-restrained system would vanish if the magnitude of the negative stiffness were sufficiently large. These results may be summarized by the following theorems:

THEOREM 1 (a).–

> The addition of one restraint with positive or negative stiffness to an n-degree of freedom system where n>1, results in a system, for which there exist at least (n-1) natural modes and frequencies.

THEOREM 1 (b).–

> As the magnitude of the stiffness parameter approaches infinity, the natural frequencies and modes of the modified restrained system would asymptotically approach those of an (n-1) degree of freedom system, which is obtained from the original system by the addition of an appropriate constraint (corresponding to the spring restraint).

7.2.2. *The natural frequencies of an n DOF system with h additional positive or negative restraints*

We can now seek to generalize the above theorems for a system with h additional restraints whose stiffnesses may be negative or positive. Since the proof is by induction, it is necessary to state the general theorems first.

THEOREM 2 (a).–

> If h restraints of positive or negative stiffness are added to an n degree of freedom system (A) where $h<n$, then for the resulting system (A_h), there exist at least (n-h) natural frequencies and modes.

THEOREM 2 (b).–

> Furthermore, as the h stiffness parameters approach infinity, the (n-h) natural frequencies and modes of System A_h would asymptotically approach those of the n degree of freedom system subject to h appropriate constraints ($\tilde{A}_h$).

Proof by Mathematical Induction

Let us denote the n degree of freedom system subject to r restraints (where $n>r>1$) having positive or negative stiffness values by A_r.

If theorem 2 (a) and (b) are true for A_r then:

there will exist (n-r) natural frequencies and modes for A_r; Statement 1 (a)

and

as the r stiffness parameters approach infinity, the natural frequencies and modes of System A_r would asymptotically approach that of the n degree of freedom system subject to r appropriate constraints ($\tilde{A}_r$). Statement 1 (b)

Applying theorem 1 (a) to system A_r, and using the Statement 1 (a), we can state that adding one more spring restraint to A_r will result in a new system A_{r+1} for which there exist (n-r-1) natural frequencies and modes.

i.e. theorem 2 (a) is true for $h = r+1$ if it is true for $h = r$. Statement 2 (a)

Applying theorem 1 (b) to $\tilde{A}_r$ with an extra restraint, we can state that as the magnitude of the stiffness of the $(r+1)^{\text{th}}$ restraint (newly added) approaches infinity, the resulting system frequencies and modes would approach that of $\tilde{A}_{r+1}$. From

Statement 1 (b), as the stiffness parameters for the r restraints approach infinity, the frequencies and modes of A_r would approach those of $\tilde{A}_r$. Therefore if the magnitude of stiffness of all r+1 restraints were to approach infinity, the natural frequencies and modes of A_{r+1} would asymptotically approach that of the n degree of freedom system subject to r+1 appropriate constraints ($\tilde{A}_{r+1}$).

i.e. theorem 2 (b) is true for $h = r+1$, if it is true for $h = r$. Statement 2 (b)

From theorem 1 (a) and (b), theorems 2(a) and (b) are true for h = 1. From Statements 2 (a) and (b) they are true for $h = r+1$ if they are true for $h = r$.

Hence by induction, theorem 2 (a) and theorem 2 (b) are true for any $h > 0$.

7.3. Applicability of theorems (1) and (2) for continuous systems

The above arguments hold for continuous systems, with the exception of any reference to the highest mode and the highest mode number n. For continuous systems, since a highest mode does not exist, the condition that $m < n$, does not apply, and when using positive stiffness values for additional springs, all natural frequencies and modes will be bounded on both sides by the natural frequencies of the corresponding constrained systems. This means equation [7.19] does not hold as n (highest mode number) does not exist, and equations [7.20]–[7.23] are true for all $m > 0$.

However, when finding the natural frequencies using the Rayleigh–Ritz procedure, continuous systems are in fact discretized and the highest mode number n corresponds to the number of terms used in the Rayleigh–Ritz formulation for the displacement.

8

Admissible Functions

In Chapter 5, we have seen that the penalty method has made the Rayleigh–Ritz method more versatile, as the admissible functions do not need to individually satisfy essential (geometric) boundary conditions or continuity conditions. However, it is well known [OOS 95] that some functions (such as polynomials) are prone to numerical instability and ill-conditioning while others (for example, sinusoidal functions in modeling clamped-clamped beams and plates) are slow to converge [DOW 84]. In this chapter we present a set of functions that have been found to be ideally suited to the vibration and stability analysis of a large set of structural systems including bars, beams, rectangular plates, shells of rectangular planform and three-dimensional (3D) bodies.

8.1. Choosing the best functions

Typical approaches to build a set of admissible functions that model the displacement of a beam to be used in the Rayleigh–Ritz method (RRM) include the use of simple polynomials, which converge fast, but are prone to ill-conditioning. To improve the accuracy of the solution, the Gram–Schmidt process as presented by Bhat [BHA 85] can be used to assure orthogonality between functions. This approach gives excellent results for plates involving free edges, as shown in [YUA 92]. The Gram–Schmidt procedure has raised some criticism by Brown and Stone [BRO 97] stating that the convergence of a vibration problem is independent of the selection of the set of admissible functions (no need for orthogonal polynomials) and that it depends only on the degree of the polynomial represented in the set. Brown and Stone [BRO 97] also stated that for plate problems, orthogonality of the functions should be targeted only on the second derivative of the functions and also recognized that special polynomials are only needed if higher order polynomials are included in the set of admissible functions. This is to make the set of functions more stable with respect to inversion and the extraction of eigenvalues of the resulting stiffness and mass matrices; although in [LI 04] Li reported that even when orthogonal polynomials are used in the RRM, the higher order polynomials become numerically unstable due to round-off errors.

Transcendental functions also have some disadvantages. For instance, Li and Daniels [LI 02] show that certain sets of admissible functions built by trigonometric functions have limitations converging when penalty parameters are included in the solution. Sets of functions using trigonometric and hyperbolic functions are very complex and are likely to become numerically unstable when several terms are used in the solution.

The best choice of functions seems to be the combination of polynomials and trigonometric functions. In publications by Li [LI 02, LI 00] and by Zhou [ZHO 96], they have shown that when polynomials and trigonometric functions are used to build sets of admissible functions, the solutions have a fast convergence rate and results are also accurate for higher modes. The approaches by Li and Zhou are similar to the one presented here except that in this work the set of functions was obtained without solving for boundary conditions. The present approach does not include an orthonormalization process.

It is important to remember that high order polynomials are the cause of numerical instabilities and ill-conditioning. Thus to keep the solution as simple as possible and free of numerical problems the minimum number of polynomial functions with the lowest order possible are included in the proposed set of admissible functions presented in this work.

8.2. Strategy for choosing the functions

Armed with the knowledge that geometric supports and continuity conditions can be enforced with penalty terms, the strategy for finding the best set of admissible functions is to consider finding a set of functions that can best model a structure in a totally unconstrained condition. Numerical experiments on a broad range of structural elements and assemblies including beams, bars, plates, shells and solids have confirmed that if a set of functions give converging and reliable results for an unconstrained system, then the results for any structure obtained by the addition of constraints can also be obtained using the same functions satisfactorily by incorporating appropriate penalty terms in the analysis. To explain the choice of functions we take an intuitive approach, but bear in mind the following requirements [OOS 95]:

a) the set of functions must be complete in energy form (all modes of vibration must be represented and no modes must be missing);

b) the set of functions must be linearly independent;

c) the functions must have derivatives at least up to half of the order of the partial differential equation.

Requirement (c) may also be considered as equivalent to the requirement that the functions must have all the derivatives in the energy terms. Now we will develop the admissible functions for an unconstrained (Free–Free) Euler–Bernoulli beam.

8.3. Admissible functions for an Euler–Bernoulli beam

As the governing partial differential equation is of the fourth order, we will need to have second order derivatives which appear in the potential energy term giving rise to the stiffness matrix. As noted earlier, we will consider combining trigonometric and simple polynomials. Sine functions can represent modes of simple supported beams exactly, as they constrain the displacement at both ends of the structure, while rotation is allowed. Similarly, cosine functions can represent the modes of sliding-sliding beams exactly, as they constrain rotation and allow translation at both ends. Thus a question arises: sine series or cosine series? The answer is found in a paper by Budiansky and Hu [BUD 46] who used cosine series in a RRM analysis together with the Lagrangian multiplier method to constrain edges of a plate. Budiansky and Hu show that the rate of convergence was faster when a cosine series was used together with translational constraints to model clamped conditions compared to the use of sine series with rotational constraints. Thus the cosine series is chosen over the sine series. The cosine series is defined as:

$$\cos\left(\frac{i\pi x}{L}\right), \text{ for } i = 0,1,2\ldots n \qquad [8.1]$$

where x is the axial coordinate of the beam, L is the beam length and n is the number of terms included in the set of admissible functions. Now, the attention must be on relaxing the sliding conditions of the cosine series making sure that the rigid-body modes are included in the set of admissible functions. Thus each of the free conditions of the beam (rotations and translations at each end) must be satisfied by at least one function. Here it is also worth noting that the cosine series includes a unit function for $i=0$. Thus the free translations are already satisfied as expected knowing that the cosine series model guiding-guiding conditions exactly. The rotational degrees of freedom of the beam are relaxed adding a linear term and a square term. Thus, the set of admissible functions $\varphi_i(x)$ developed here are

$$\varphi_i(x)=1, \text{ for } i=1 \qquad [8.2a]$$

$$\varphi_i(x)=\left(\frac{x}{L}\right), \text{ for } i=2 \qquad [8.2b]$$

$$\varphi_i(x)=\left(\frac{x}{L}\right)^2, \text{ for } i=3 \qquad [8.2c]$$

$$\varphi_i(x)=\cos\frac{(i-3)\pi x}{L}, \text{ for } i=4,5,\ldots n \qquad [8.2d]$$

In this work, functions in equations [8.2] will be used in the RRM to model the transverse deflection of the beam given by

$$w(x,t)=W(x)\sin(\omega t), \qquad [8.3a]$$

where $W(x)$ is the amplitude of the deflection of the neutral axis of the beam defined as

$$W(x)=\sum_{i=1}^{n}c_i\varphi_i(x), \qquad [8.3b]$$

where c_i are arbitrary coefficients.

Fourier series are orthogonal functions with respect to each other when integrated to the full span [0 to L] as shown below [SZI 04]:

$$\int_0^L \cos\frac{i\pi x}{L}\cos\frac{j\pi x}{L}dx=\begin{cases}0 & \text{for } i\neq j\\ L/2 & \text{for } i=j\end{cases}, \qquad [8.4]$$

Furthermore the first and second derivatives of orthogonal functions are also orthogonal [SZI 04]. This property is very useful to reduce the effort in obtaining the terms of the elastic stiffness, geometrical stiffness and mass matrices of structural elements presented later such as beams and plates and shells. Sets of orthogonal functions used in the RRM produce diagonal mass and stiffness. However, in [MUK 08] Mukhopadhyay stated that when a good set of admissible functions is chosen, the off-diagonal terms are small. In the present work, the mass matrix of a beam has off-diagonal terms only in the rows and columns corresponding to the linear and square functions, because they are not orthogonal with any other functions of the set of admissible functions. The absolute value of these off-diagonal terms of the mass matrix decreases as the number of terms in the set of admissible functions increases, starting with the values of the fourth admissible function. However the stiffness matrix of a beam is a diagonal matrix, as orthogonality was kept for the second derivatives as suggested by Brown and Stone [BRO 97]. This would not be the case if another simple polynomial function of an order higher than two is included in the set of admissible functions. Thus, the stiffness matrix of a beam derived with the present set of admissible functions results in a diagonal matrix, although the values of the first two terms in the main diagonal are zero. For all other structural elements the stiffness, geometrical stiffness and mass matrices are sparse. Furthermore, the set of admissible functions does not produce ill-conditioning due to the number of terms used in the series as shown by Monterrubio in [MON 09, MON 12a] and [MON 12b]. This is a very important characteristic to be able to define constraints of complex geometry and/or interconnect structural elements

We will now further clarify the role of the functions in satisfying the boundary conditions. The following inequalities show that the selected set of functions permit non-zero displacement and translation at both ends as required for a free-free beam.

$\varphi_i(0) \neq 0$ This condition is satisfied by equations [8.2a], [8.2d],

$\varphi_i(L) \neq 0$ All functions included in the set defined in equations [8.2a], [8.2d] satisfy this condition,

$\left.\dfrac{\partial \varphi_i}{\partial x}\right|_{x=0} \neq 0$ This condition is only satisfied by the linear term defined in equation [8.2b] and

$\left.\dfrac{\partial \varphi_i}{\partial x}\right|_{x=L} \neq 0$ This condition is satisfied by the linear and square terms defined in equations [8.2b], [8.2c].

The argument above shows that the proposed set of admissible functions is a complete set, which models the deflection of a free-free beam.

We will now show why these functions should converge.

8.4. Proof of convergence

For the RRM with undetermined coefficients in the above form to converge to the modes of a free–free beam, the set of admissible functions must allow [BAT 96]:

1) rigid body motion without inducing any strain;

2) the functions must permit constant strain to be achieved.

Following the same steps as in building the set of admissible functions, we can start with the cosine series functions $f_i(x) = \cos(i\pi x / L)$ which are the natural modes of a sliding-sliding beam. Thus, using the cosine series above in the RRM to solve a vibration problem, gives the exact values of the natural frequencies and modes of vibration. Here it is worth recalling that the results of a Rayleigh–Ritz analysis are as close to the exact solution as the accuracy with which the modes of vibration are modeled in the set of admissible functions. As noted before, the cosine series includes the unit function which models the translational rigid-body of the beam.

Now consider the functions $g(x) = (x / L)$ and $h(x) = (x / L)^2$ whose first derivatives $g'(x) \neq 0$ for $x = 0$ and $h'(x) \neq 0$ for $x = L$. The inclusion of these two functions allows the second rigid-body mode of the beam satisfying point a) above. Furthermore, the function $(x / L)^2$ allows constant curvature to be represented, which for an Euler–Bernoulli beam implies constant strain along the axis satisfying point b) above.

It should also be clear that when using $f(x)$, $g(x)$ and $h(x)$ to build $\varphi_i(x)$ which is included in equation [8.3b].

$$W(x) = \sum_{i=1}^{n} c_i \varphi_i(x), \tag{8.3b}$$

The transverse deflection of the neutral axis of a Bernoulli beam can take any smooth shape, which obviously include the guiding–guiding modes and as the translational rigid-body mode is included with the unit function, while the rotational rigid-body mode is possible due to the linear and square functions which relax the constraints impose by the cosine series.

From Rayleigh's theorem of separation, minimizing the Rayleigh quotient, $\frac{V_m}{\psi_m}$ results in eigenvalues that are upper bounds to the true solution. This shows that our choice of admissible functions could form a complete and converging series to represent the natural modes of a free-free beam. This convergence test is however not a sufficient condition.

9

Natural Frequencies and Modes of Beams

9.1. Introduction

In the previous chapter we have identified a convenient set of admissible functions for use in the RRM to find natural frequencies and modes of common structural elements. The simplest of these is an Euler–Bernoulli beam. In this chapter, we will see how the RRM can be employed to determine the natural frequencies and modes of Euler–Bernoulli beams with all combinations of standard boundary conditions (F – free, S – simply supported, C – clamped and G – guided). The solution procedure for the minimization is presented below as a method to obtain the mass and stiffness matrices of an unconstrained beam, as well as the penalty matrices that define constraints of the beam. For convenience, the system matrices (mass and stiffness matrices) and results for the circular natural frequencies are all presented in non-dimensional form. Results for the frequency parameters are compared with results published in the literature. Modes of vibration are also included.

9.2. Theoretical derivations of the eigenvalue problems

In this chapter, the set of admissible functions $\varphi_i(x)$ is taken as a combination of a complete quadratic polynomial and a cosine series, described in equation [8.2], that is:

$$\varphi_i(x) = 1 \quad i = 1 \qquad [8.2a]$$

$$\varphi_i(x) = \frac{x}{L} \text{ if } i = 2 \qquad [8.2b]$$

$$\varphi_i(x) = \left(\frac{x}{L}\right)^2 \text{ if } i = 3 \qquad [8.2c]$$

$$\varphi_i(x) = \cos\frac{(i-3)\pi x}{L} \text{ if } i = 4,5,6,\ldots,n \qquad [8.2d]$$

where x is the coordinate along the beam axis, L is the beam length and n is the maximum number of terms included in the set of admissible functions.

The deflection of the beam $w(x)$ is defined as:

$$w(x) = W(x)\cos(\omega t), \qquad [9.1]$$

where ω is the frequency of oscillation, t is time and $W(x)$ is the deflected shape of the neutral line of the beam that can be expressed in terms of the set of admissible functions as

$$W(x) = \sum_{i=1}^{n} a_i\varphi_i = a_1 + a_2x + a_3x^2 + \sum_{i=4}^{n} a_i \cos\frac{(i-3)\pi x}{L} \qquad [9.2a]$$

or in non-dimensional form:

$$\hat{W}(\xi) = \sum_{i=1}^{n} a_i\hat{\varphi}_i = a_1 + a_2\xi + a_3\xi^2 + \sum_{i=4}^{n} a_i \cos(i-3)\pi\xi, \qquad [9.2b]$$

where $\xi = x/L$ is the non-dimensional coordinate and a_i are arbitrary coefficients.

Thus,

$$\hat{\varphi}_i(\xi) = 1 \text{ if } i = 1 \qquad [9.3a]$$

$$\hat{\varphi}_i(\xi) = \xi \text{ if } i = 2 \qquad [9.3b]$$

$$\hat{\varphi}_i(\xi) = \xi^2 \text{ if } i = 3 \qquad [9.3c]$$

$$\hat{\varphi}_i(\xi) = \cos(i-3)\pi\xi \text{ if } i = 4,5,6n,\ldots,n \qquad [9.3d]$$

These functions combined with penalty parameters allow us to model beams with all possible combinations of boundary conditions: the classical free (F), simply supported (S) and clamped (C), as well as sliding (G) conditions.

9.3. Derivation of the eigenvalue problem for beams

Consider an Euler–Bernoulli beam of constant thickness h with elastic translational and rotational supports as shown in Figure 9.1.

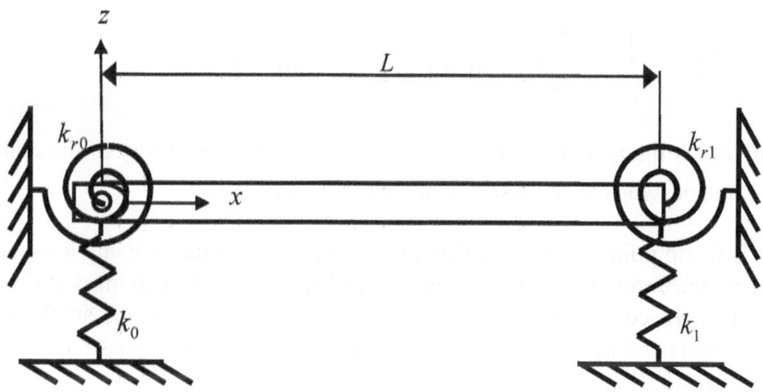

Figure 9.1. *Beam of length L with translational springs (k_0 and k_1) and rotational springs (k_{r0} and k_{r1})*

The expressions for the potential and kinetic energy terms needed to compute the natural frequencies and modes of vibration were given previously including Chapter 6. Thus for all beam cases with constraints at one or both ends of a beam, springs are located at $x_0 = 0$ and $x_1 = L$ (or $\xi_0 = 0$ and $\xi_1 = L$ in non-dimensional coordinates) if required. Then the maximum potential energy $V_{\max}$ of the beam in Figure 9.1 and the strain energy of the translational springs with stiffness coefficients k_0 and k_1 and rotational springs with coefficient k_{r0} and k_{r1} are

$$V_{\max} = \frac{EI}{2}\int_{x_0=0}^{x_1=L}\left(\frac{\partial^2 W}{\partial x^2}\right)^2 dx + \frac{k_0}{2}W^2\Big|_{x_0=0} + \frac{k_1}{2}W^2\Big|_{x_1=L} + \frac{k_{r0}}{2}\left(\frac{\partial W}{\partial x}\right)^2\Big|_{x_0=0} + \frac{k_{r0}}{2}\left(\frac{\partial W}{\partial x}\right)^2\Big|_{x_1=L}, \quad [9.4]$$

where EI is the flexural rigidity of the beam (E is the Young's modulus and I is the second moment of area).

Here the springs may be used to enforce full constraints by using very large magnitudes for the stiffness coefficients. If masses and moments of inertia m_0, m_1, I_0 and I_1 respectively substitute the translational and rotational springs k_0, k_1, k_{r0} and k_{r1} in Figure 9.1, the maximum kinetic energy T_{max} of the system becomes

$$T_{max} = \frac{\omega^2 \rho A}{2} \int_{x_0=0}^{x_1=L} W^2 dx + \frac{\omega^2 m_0}{2} W^2 \Big|_{x_0=0} + \frac{\omega^2 m_1}{2} W^2 \Big|_{x_1=L} + \frac{\omega^2 I_0}{2} \left(\frac{\partial W}{\partial x} \right)^2 \Bigg|_{x=0} + \frac{\omega^2 I_1}{2} \left(\frac{\partial W}{\partial x} \right)^2 \Bigg|_{x_1=L}, \quad [9.5]$$

where ω^2 is the natural circular frequency and ρ is the density of the material of the beam and A is the cross-sectional area of the beam.

It is worth pointing out here that any number of constraints at any location of the beam can be enforced by locating an artificial spring or artificial mass at the desired location of the constraint. Simultaneous use of artificial springs and artificial masses to model constraints known as the "bipenalty method" is also possible as shown in [ASK 10] and [ILA 12].

The maximum kinetic energy function Γ_{max} to be minimized is then given by:

$$\Gamma_{max} = T_{max} / \omega^2, \quad [9.6]$$

For convenience let us rewrite equations [9.4]–[9.6] in non-dimensional form. The non-dimensional strain energy is obtained by dividing the strain energy of the system given in equation [9.4] by the term EI / L^3, which is a measure of the flexural stiffness of the beam, using the non-dimensional deflected shape of the neutral line of the beam $\hat{W}$ and a non-dimensional stiffness penalty parameter $\hat{p}_s$ defined as

$$\hat{p}_s = \frac{k_i L^3}{EI} = \frac{k_{ri} L}{EI}, \quad [9.7]$$

Thus, the non-dimensional strain energy $\hat{V}_{\max}$ of the system is:

$$\hat{V}_{\max} = \frac{1}{2}\int_{\xi_0=0}^{\xi_1=1}\left(\frac{\partial^2 \hat{W}}{\partial \xi^2}\right)^2 d\xi + \frac{\hat{p}_s}{2}\hat{W}^2\Big|_{\xi_0=0} + \frac{\hat{p}_s}{2}\hat{W}^2\Big|_{\xi_1=1} + \frac{\hat{p}_s}{2}\left(\frac{\partial \hat{W}}{\partial \xi}\right)^2\Bigg|_{\xi_0=0} + \frac{\hat{p}_s}{2}\left(\frac{\partial \hat{W}}{\partial \xi}\right)^2\Bigg|_{\xi_1=1}, \qquad [9.8]$$

Similarly, the non-dimensional kinetic energy of the system is obtained by dividing the kinetic energy of the system given in equation [9.5] by the mass of the beam $m = \rho AL$, using the non-dimensional deflection shape of the neutral line of the beam $\hat{W}$ and a non-dimensional inertial penalty parameter $\hat{p}_m$ defined as:

$$\hat{p}_m = \frac{m_i}{\rho AL} = \frac{I_i}{\rho AL^3}, \qquad [9.9]$$

Thus, the non-dimensional kinetic energy of the system is:

$$\hat{T}_{\max} = \frac{\omega^2}{2}\int_{\xi_0=0}^{\xi_1=1}\hat{W}^2 d\xi + \frac{\omega^2 \hat{p}_m}{2}\hat{W}^2\Big|_{\xi_0=0} + \frac{\omega^2 \hat{p}_m}{2}\hat{W}^2\Big|_{\xi_1=1} + \frac{\omega^2 \hat{p}_m}{2}\left(\frac{\partial \hat{W}}{\partial \xi}\right)^2\Bigg|_{\xi=0} + \frac{\omega^2 \hat{p}_m}{2}\left(\frac{\partial \hat{W}}{\partial \xi}\right)^2\Bigg|_{\xi_1=1}, \qquad [9.10]$$

It is worth pointing out that artificial masses and/or artificial springs should be added to the system to model a constraint in the system.

The beam is assumed to be a conservative system; by conservation of energy it then follows that:

$$V_{\max} - T_{\max} = 0 \quad \text{or} \quad \hat{V}_{\max} - \hat{T}_{\max} = 0, \qquad [9.11a, 9.11b]$$

The stiffness and mass matrices $\mathbf{K}$ and $\mathbf{M}$ are obtained by applying the Rayleigh–Ritz minimization to the conservative system which yields,

$$\frac{V_{\max}}{\partial a_i} - \omega^2 \frac{\Gamma_{\max}}{\partial a_i} = 0 \quad \text{or} \quad \mathbf{K} - \{\omega^2\}\mathbf{M} = \{0\} \qquad [9.12a, 9.12b]$$

or in non-dimensional form,

$$\frac{\hat{V}_{\max}}{\partial a_i} - \lambda^4 \frac{\hat{\Gamma}_{\max}}{\partial a_i} = 0 \quad \text{or} \quad \hat{\mathbf{K}} - \{\omega^2\}\hat{\mathbf{M}} = \{0\} \qquad [9.12c, 9.12d]$$

Solving the eigenproblem of equation [9.12d] allows us to obtain the non-dimensional frequency parameter λ defined as [ILA 99]:

$$\lambda_i = \sqrt[4]{\frac{\rho A L^4 \omega^2}{EI}} \qquad [9.13]$$

9.4. Building the stiffness, mass matrices and penalty matrices

In general, when using the RRM, the stiffness matrix and the mass matrix of structural elements are obtained by carrying out the Rayleigh-Ritz minimization process. Specifically, the stiffness matrix of a beam is obtained from integrals of all combinations of the second derivatives of the admissible functions, while the mass matrix of a beam is obtained by carrying out the integrals of all the combinations of the admissible functions (as shown below). Stiffness and mass matrices of other elements involve integrations of different combinations of the admissible functions and their derivatives.

The non-dimensional potential energy of the unconstrained beam and artificial springs are given in equation [9.8]. Similarly, the non-dimensional kinetic energy of the beam and artificial masses and moments of inertia are given in equation [9.10]. For clarity, in this work the stiffness and mass matrices of the beam are obtained first from the Rayleigh–Ritz minimization of the first term of equation [9.8] and the first term of equation [9.10], respectively.

Typically, the following nomenclature is used to define the terms of the stiffness and mass matrices:

$$E_{ij}^{(r,s)} = \int_{\xi_0=0}^{\xi_1=1} \left(\mathrm{d}^r \hat{\varphi}_i(\xi)/\mathrm{d}\xi^r\right)\left(\mathrm{d}^s \hat{\varphi}_j(\xi)/\mathrm{d}\xi^s\right)\,\mathrm{d}\xi \qquad [9.14]$$

Penalty terms to be added to the stiffness and mass matrices are obtained from the minimization of the terms containing the non-dimensional penalty parameter $\hat{p}_s$ in equation [9.8] and $\hat{p}_m$ in equation [9.10]. Remembering that the minimization is carried out to the maximum kinetic energy function, $\Gamma_{\max} = T_{\max}/\omega^2$, it should be clear that for equal values of the non-dimensional penalty parameters $\hat{p}_s = \hat{p}_m$, the penalty matrices that are added to the stiffness matrix and/or to the mass matrix are identical. In other words, the only difference in the penalty matrices is in the value of the penalty parameter. For this reason, and because no specific examples using artificial springs and artificial masses are given in this chapter, the penalty parameters will be defined as $\hat{p}$. Translational constraints are modeled by penalty matrices built by the combination of all admissible functions evaluated at the location of the constraints, whilst rotational constraints are modeled by penalty matrices built by the permutation of the first derivatives of all admissible functions evaluated at the location of the constraints.

Penalty matrices for plates, shells and 3D bodies that add a constraint along an edge can be defined by an integral as shown in the following chapters.

9.4.1. *Terms* $\hat{K}_{ij}$ *of the non-dimensional stiffness matrix* $\hat{\mathbf{K}}$

As shown in earlier chapters, the terms of the stiffness matrix of a beam are the result of the Rayleigh–Ritz minimization of its strain energy, whose non-dimensional equivalent is:

$$\hat{K}_{ij} = \frac{\partial^2}{\partial a_i a_j}\left(\frac{1}{2}\int_{\xi_0=0}^{\xi_1=1}\left(\frac{\partial^2 \hat{W}}{\partial \xi^2}\right)^2 d\xi\right) = \int_{\xi_0=0}^{\xi_1=1} \hat{\varphi}_i'' \hat{\varphi}_j'' d\xi\,, \qquad [9.15]$$

Note that the present procedure is valid for constant flexural rigidity of Euler–Bernoulli beams EI. Otherwise, if the flexural rigidity can be described as a function of the non-dimensional co-ordinate ξ, this function has to be included in the integral defined in equation [9.15].

Table 9.1 shows the non-dimensional admissible functions and their first and second derivatives.

	$\hat{\varphi}_i$	$\hat{\varphi}_i'$	$\hat{\varphi}_i''$
$i=1$	1	0	0
$i=2$	ξ	1	0
$i=3$	ξ^2	2ξ	2
$i\geq 4$	$\cos\left((i-3)\pi\xi\right)$	$-(i-3)\pi\sin\left((i-3)\pi\xi\right)$	$-(i-3)^2\pi^2\cos\left((i-3)\pi\xi\right)$

Table 9.1. *First four admissible functions* $\hat{\varphi}_i$ *and their corresponding first derivatives* $\hat{\varphi}_i'$ *and second derivatives* $\hat{\varphi}_i''$

A symmetric matrix showing the product of the second derivatives of the admissible functions for different values of the subscripts i and j that appear in equation [9.15] is shown in Table 9.2.

$i \backslash j$	$\hat{\varphi}''_{j=1}$	$\hat{\varphi}''_{j=2}$	$\hat{\varphi}''_{j=3}$	$\hat{\varphi}''_{j\geq 4}$
$\hat{\varphi}''_{i=1}$	0	0	0	0
$\hat{\varphi}''_{i=2}$		0	0	0
$\hat{\varphi}''_{i=3}$	SYMM		4	$-2(j-3)^2\pi^2\cos((j-3)\pi\xi)$
$\hat{\varphi}''_{i\geq 4}$				$(i-3)^2(j-3)^2\pi^4\cos((i-3)\pi\xi)\cos((j-3)\pi\xi)$

Table 9.2. *Products of the second derivatives of the set of admissible functions*

The last step to calculate the non-dimensional stiffness matrix of a beam is to carry out the integration of the terms in Table 9.2 with respect to the non-dimensional coordinate ξ with limits of integration 0 and 1 as defined in equation [9.15].

Before carrying out the integrals, it is worth noting that:

– the second derivatives of the first two functions are zero. Thus the first two rows and columns of the products of the functions shown in Table 9.2 are full of zeroes;

– using the parametric description of the terms as presented in Tables 9.1 and 9.2 it is only necessary to describe the matrices up to the fourth column and fourth row. Adding a fifth term in the set of admissible functions gives the same terms defined in the fourth column and fourth row but values of i and j need to be changed as appropriate;

– as expected the terms in Table 9.2 are symmetric about the main diagonal.

Thus, only the following three integrals are needed to define the stiffness matrix:

1) for $i=j=3$

$$\hat{K}_{33} = \int_{\xi=0}^{\xi=1} 4d\xi = \int_{\xi=0}^{\xi=1} 4d\xi = 4$$

2) for $i=3$ and $j\geq 4$

$$\hat{K}_{i=3,j\geq 4} = \int_{\xi=0}^{\xi=1} -2(i-3)^2\pi^2\cos((i-3)\pi\xi)d\xi$$

Taking the constant terms out of the integral gives:

$$\hat{K}_{i=3,j\geq 4} = 2(j-3)^2 \pi^2 \int_{\xi=0}^{\xi=1} -\cos((j-3)\pi\xi)\, d\xi$$

and carrying out the integration:

$$\hat{K}_{i=3,j\geq 4} = \left(2(j-3)^2 \pi^2\right)\left(\left(\frac{1}{(j-3)\pi}\right)\sin((j-3)\pi\xi)\Bigg|_{\xi=0}^{\xi=1}\right)$$

Knowing that $\sin(n\pi\xi) = 0$, where n is an integer, reduces the above equation to $\hat{K}_{i=3,j\geq 4} = 0$. This makes the stiffness matrix a diagonal matrix as all off-diagonal terms are now equal to zero.

3) for $i = j,\ j \geq 4$

$$K_{i=j,j\geq 4} = \int_{\xi=0}^{\xi=1} (i-3)^2 (j-3)^2 \pi^4 \cos((i-3)\pi\xi)\cos((j-3)\pi\xi)\, d\xi$$

Taking the constant terms out of the integral gives:

$$K_{i=j,j\geq 4} = (i-3)^2 (j-3)^2 \pi^4 \int_{\xi=0}^{\xi=1} \cos((i-3)\pi\xi)\cos((j-3)\pi\xi)\, d\xi$$

Using the following relationship given by Szilard in [9.30]

$$\int_{\xi=0}^{\xi=1} \cos(i\pi\xi)\cos(j\pi\xi)\, d\xi = \begin{cases} 0 & \text{for} \quad i \neq j \\ 1/2 & \text{for} \quad i = j \end{cases} \quad \text{gives}$$

$$K_{i=j,j\geq 4} = \frac{1}{2}(i-3)^2 (j-3)^2 \pi^4 = \frac{1}{2}(i-3)^4 \pi^4$$

Thus, the non-dimensional stiffness matrix $\hat{\mathbf{K}}$ of an Euler–Bernoulli beam computed using the set of admissible functions defined in equation [9.2] in the RRM is:

$$\hat{\mathbf{K}} = \begin{array}{c} \\ \\ \\ \\ \\ \end{array}\begin{array}{cccc} \hat{\varphi}''_{j=1} & \hat{\varphi}''_{j=2} & \hat{\varphi}''_{j=3} & \hat{\varphi}''_{j=i\geq 4} \\ \left[\begin{array}{c} 0 \\ \\ \\ \text{Sym} \end{array}\right. & \begin{array}{c} 0 \\ 0 \\ \\ \\ \end{array} & \begin{array}{c} 0 \\ 0 \\ 4 \\ \\ \end{array} & \left.\begin{array}{c} 0 \\ 0 \\ 0 \\ \frac{1}{2}(i-3)^4\pi^4 \end{array}\right] \end{array} \begin{array}{c} \\ \hat{\varphi}''_{i=1} \\ \hat{\varphi}''_{i=2} \\ \hat{\varphi}''_{i=3} \\ \hat{\varphi}''_{i=j\geq 4} \end{array} \quad [9.16a]$$

The corresponding dimensionalized results for the stiffness matrix $\mathbf{K}$ can be obtained multiplying $\hat{\mathbf{K}}$ by the factor EI / L^3. Thus:

$$\mathbf{K} = \hat{\mathbf{K}} EI / L^3, \quad [9.16b]$$

Using the nomenclature defined in equation [9.14] the terms of the non-dimensional stiffness matrix of a beam can be defined as:

$$\hat{K}_{ij} = E_{i,j}^{(2,2)}, \quad [9.17]$$

9.4.2. *Terms $\hat{M}_{ij}$ of the non-dimensional mass matrix $\hat{\mathbf{M}}$*

Following the same procedure to build the mass matrix of an Euler–Bernoulli beam, the product of the admissible functions is given in Table 9.3.

As shown in earlier chapters, the terms of the mass matrix of a beam are the result of the Rayleigh–Ritz minimization of its kinetic energy function, whose non-dimensional equivalent is:

$$M_{ij} = \frac{\partial^2}{\partial a_i \partial a_j}\left(\frac{1}{2}\int_{\xi_0=0}^{\xi_1=1} \hat{W}^2 d\xi\right) = \frac{1}{2}\int_{\xi_0=0}^{\xi_1=1} \hat{\varphi}_i \hat{\varphi}_j d\xi \quad [9.18]$$

$i \backslash j$	$\hat{\varphi}_{j=1}$	$\hat{\varphi}_{j=2}$	$\hat{\varphi}_{j=3}$	$\hat{\varphi}_{j\geq 4}$
$\hat{\varphi}_{i=1}$	1	ξ	ξ^2	$\cos((j-3)\pi\xi)$
$\hat{\varphi}_{i=2}$		ξ^2	ξ^3	$\xi\cos((j-3)\pi\xi)$
$\hat{\varphi}_{i=3}$	Sym		ξ^4	$\xi^2\cos((j-3)\pi\xi)$
$\hat{\varphi}_{i\geq 4}$				$\cos((i-3)\pi\xi)\ \cos((j-3)\pi\xi)$

Table 9.3. *Products of the set of admissible functions*

All the integrals needed to compute the terms of the non-dimensional mass matrix are very simple and easy to compute by hand. The first three functions are simple polynomials and the resulting terms are defined by simple integrals as follows:

$$M_{11} = \int_{\xi_0=0}^{\xi_1=1} 1 d\xi = 1$$

$$M_{12} = \int_{\xi_0=0}^{\xi_1=1} \xi d\xi = \frac{1}{2}$$

$$M_{13} = M_{22} = \int_{\xi_0=0}^{\xi_1=1} \xi^2 d\xi = \frac{1}{3}$$

$$M_{23} = \int_{\xi_0=0}^{\xi_1=1} \xi^3 d\xi = \frac{1}{4}$$

$$M_{33} = \int_{\xi_0=0}^{\xi_1=1} \xi^3 d\xi = \frac{1}{5}$$

The integration of the products in the fourth column in Table 9.3 for $\hat{\varphi}_{j\geq 4}$ are slightly more challenging. The first term in the fourth column of Table 9.3 is the integral of a cosine function, which gives a sine function. As shown earlier, sine functions evaluated at zero and at integer numbers of π are equal to zero. The second term of the fourth column is:

$$M_{24} = \int_{\xi_0=0}^{\xi_1=1} \xi \cos\left(\left(j-3\right)\pi\xi\right) d\xi = 0,$$

which is computed using integration by parts $\int udv = uv - \int vdu$. In this case $u = \xi$ and $dv = \cos\left(\left(j-3\right)\pi\xi\right)d\xi$ and $du = d\xi$, $v = \frac{1}{\left(j-3\right)\pi}\sin\left(\left(j-3\right)\pi\xi\right)$

$$M_{24} = \frac{\xi}{\left(j-3\right)\pi}\sin\left(\left(j-3\right)\pi\xi\right)\Bigg|_{\xi_1=0}^{\xi_1=1} - \int_{\xi_1=0}^{\xi_1=1} \frac{1}{\left(j-3\right)\pi}\sin\left(\left(j-3\right)\pi\xi\right)d\xi$$

The first term is zero because it contains a sine function and carrying out the integral of the second term gives:

$$M_{24} = \frac{1}{\left(j-3\right)^2\pi^2}\cos\left(\left(j-3\right)\pi\xi\right)\Bigg|_{\xi_1=0}^{\xi_1=1} = \frac{1}{\left(j-3\right)^2\pi^2}\left(\cos\left(\left(j-3\right)\pi\right)-1\right)$$

The third term of the fourth column is:

$$M_{34} = \int_{\xi_0=0}^{\xi_1=1} \xi^2 \cos\left(\left(j-3\right)\pi\xi\right) d\xi \,,$$

In this case it is necessary to perform integration by parts twice. The result of the integral is:

$$M_{34} = \frac{2}{\left(j-3\right)^2 \pi^2} \cos\left(\left(j-3\right)\pi\right)$$

The fourth term of the fourth column for terms along the main diagonal ($i = j$) is:

$$M_{44} = \int_{\xi_0=0}^{\xi_1=1} \cos\left(\left(i-3\right)\pi\xi\right) \cos\left(\left(j-3\right)\pi\xi\right) d\xi = \frac{1}{2} \quad \text{or zero if } i \neq j$$

Thus, the non-dimensional mass matrix of an Euler–Bernoulli beam computed using the non-dimensional set of admissible functions defined in equation [9.3] in the RRM is:

$$\begin{array}{c} \\ \hat{\mathbf{M}} = \end{array} \begin{array}{c} \begin{array}{cccc} \hat{\varphi}_{j=1} & \hat{\varphi}_{j=2} & \hat{\varphi}_{j=3} & \hat{\varphi}_{j\geq 4} \end{array} \\ \left[\begin{array}{cccc} 1 & \frac{1}{2} & \frac{1}{3} & 0 \\ & \frac{1}{3} & \frac{1}{4} & \frac{\cos\left(\left(j-3\right)\pi\right) -}{\left(j-3\right)^2 \pi^2} \\ & & \frac{1}{5} & \frac{2\cos\left(\left(j-3\right)\pi\right)}{\left(j-3\right)^2 \pi^2} \\ \text{Sym} & & & \frac{1}{2} \end{array} \right] \end{array} \begin{array}{c} \\ \hat{\varphi}_{i=1} \\ \hat{\varphi}_{i=2} \\ \hat{\varphi}_{i=3} \\ \varphi_{i\geq 4} \end{array} \qquad [9.19a]$$

The dimensional mass matrix $\mathbf{M}$ can be computed multiplying $\hat{\mathbf{M}}$ as shown below:

$$\mathbf{M} = \hat{\mathbf{M}} \rho A L \qquad [9.19b]$$

9.4.3. *Terms* $\hat{P}_{ij}$ *of the non-dimensional penalty matrix* $\hat{\mathbf{P}}$

Penalty matrices in this work are the result of the Rayleigh–Ritz minimization procedure of the strain energy of the artificial springs (last four terms in the potential energy equation [9.8])

$$V_{springs} = \frac{\hat{p}}{2}\hat{W}^2\bigg|_{\xi_0=0} + \frac{\hat{p}}{2}\hat{W}^2\bigg|_{\xi_1=1} + \frac{\hat{p}}{2}\left(\frac{\partial \hat{W}}{\partial \xi}\right)^2\bigg|_{\xi_0=0} + \frac{\hat{p}}{2}\left(\frac{\partial \hat{W}}{\partial \xi}\right)^2\bigg|_{\xi_1=1} \qquad [9.20]$$

or the minimization of the kinetic energy function of the artificial masses and inertia springs (last four terms in the kinetic energy equation [9.10] divided by ω^2)

$$T_{mass} = \frac{\hat{p}}{2}\hat{W}^2\bigg|_{\xi_0=0} + \frac{\hat{p}}{2}\hat{W}^2\bigg|_{\xi_1=1} + \frac{\hat{p}}{2}\left(\frac{\partial \hat{W}}{\partial \xi}\right)^2\bigg|_{\xi=0} + \frac{\hat{p}}{2}\left(\frac{\partial \hat{W}}{\partial \xi}\right)^2\bigg|_{\xi_1=1} \qquad [9.21]$$

$$\begin{array}{cccccc} & \hat{\varphi}''_{j=1} & \hat{\varphi}''_{j=2} & \hat{\varphi}''_{j=3} & \hat{\varphi}''_{j\geq 4} & \\ \hat{\mathbf{P}}_{translation} = \hat{p} \left[\right. & 1 & \xi & \xi^2 & \cos\left((j-3)\pi\xi\right) & \hat{\varphi}''_{i=1} \\ & & \xi^2 & \xi^3 & \xi\cos\left((j-3)\pi\xi\right) & \hat{\varphi}''_{i=2} \\ & & & \xi^4 & \xi^2\cos\left((j-3)\pi\xi\right) & \hat{\varphi}''_{i=3} \\ & \text{Sym} & & & \cos\left((i-3)\pi\xi\right)\cos\left((j-3)\pi\xi\right) \left. \right] & \varphi''_{i\geq 4} \end{array} \qquad [9.22]$$

The general form of a penalty matrix to model a translational constraint for an Euler–Bernoulli beam, using the set of admissible function defined in equation [9.3], is equal to the penalty parameter times a matrix built with the terms contained in Table 9.3, evaluating ξ at the location of the constraint. Thus, the general form of a penalty matrix to model a translational constraint for an Euler–Bernoulli beam using the set of admissible functions defined in equation [9.3] is equal to the penalty parameter times a matrix built with the terms contained in Table 9.3, evaluating ξ at the location of the constraint.

Then a translational constrained at $\xi = 0$ can be modeled as:

$$\hat{\mathbf{P}}_{translation} = \hat{p} \begin{bmatrix} 1 & 0 & 0 & 1 \\ & 0 & 0 & 0 \\ & & 0 & 0 \\ \text{Sym} & & & 1 \end{bmatrix} \quad \begin{matrix} \hat{\varphi}''_{j=1} & \hat{\varphi}''_{j=2} & \hat{\varphi}''_{j=3} & \hat{\varphi}''_{j\geq 4} \end{matrix} \quad \begin{matrix} \hat{\varphi}''_{i=1} \\ \hat{\varphi}''_{i=2} \\ \hat{\varphi}''_{i=3} \\ \varphi''_{i\geq 4} \end{matrix} \qquad [9.23]$$

while a translational constrained at $\xi = 1$ can be modeled using the following penalty matrix. Note that values of the terms in rows and columns $i \geq 4, j \geq 4$ are either 1 or -1.

$$\hat{\mathbf{P}}_{translation} = \hat{p} \begin{bmatrix} 1 & 1 & 1 & \cos((j-3)\pi) \\ & 1 & 1 & \cos((j-3)\pi) \\ & & 1 & \cos((j-3)\pi) \\ \text{Sym} & & & \cos((i-3)\pi)\cos((j-3)\pi) \end{bmatrix} \quad \begin{matrix} \hat{\varphi}_{j=1} & \hat{\varphi}_{j=2} & \hat{\varphi}_{j=3} & \hat{\varphi}_{j\geq 4} \end{matrix} \quad \begin{matrix} \hat{\varphi}_{i=1} \\ \hat{\varphi}_{i=2} \\ \hat{\varphi}_{i=3} \\ \varphi_{i\geq 4} \end{matrix}$$

[9.24]

Similarly, rotational constraints are generally defined by a matrix with the terms defined in Table 9.4 multiplied by a penalty parameter.

$i \backslash j$	$\hat{\varphi}'_{j=1}$	$\hat{\varphi}'_{j=2}$	$\hat{\varphi}'_{j=3}$	$\hat{\varphi}'_{j\geq 4}$
$\hat{\varphi}'_{i=1}$	0	0	0	0
$\hat{\varphi}'_{i=2}$		1	2ξ	$-(j-3)\pi\sin((j-3)\pi\xi)$
$\hat{\varphi}'_{i=3}$	Sym		$4\xi^2$	$-2\xi(j-3)\pi\sin((j-3)\pi\xi)$
$\hat{\varphi}'_{i\geq 4}$				$(i-3)\pi\sin((i-3)\pi\xi)(j-3)\pi\sin((j-3)\pi\xi)$

Table 9.4. *Products of the first derivative of the set of admissible functions*

Thus, to model a rotational constraint at $\xi = 0$ the following penalty matrix must be included in the analysis:

$$\hat{\mathbf{P}}_{translation} = \hat{p} \begin{array}{c} \begin{array}{cccc} \hat{\varphi}_{j=1} & \hat{\varphi}_{j=2} & \hat{\varphi}_{j=3} & \hat{\varphi}_{j\geq 4} \end{array} \\ \left[\begin{array}{cccc} 0 & 0 & 0 & 0 \\ & 1 & 0 & 0 \\ & & 0 & 0 \\ \text{Sym} & & & 0 \end{array} \right] \end{array} \begin{array}{c} \hat{\varphi}_{i=1} \\ \hat{\varphi}_{i=2} \\ \hat{\varphi}_{i=3} \\ \varphi_{i\geq 4} \end{array} \qquad [9.25]$$

Similarly, to model a rotational constraint at $\xi = 1$ the following penalty matrix must be included in the analysis:

$$\hat{\mathbf{P}}_{translation} = \hat{p} \begin{array}{c} \begin{array}{cccc} \hat{\varphi}_{j=1} & \hat{\varphi}_{j=2} & \hat{\varphi}_{j=3} & \hat{\varphi}_{j\geq 4} \end{array} \\ \left[\begin{array}{cccc} 0 & 0 & 0 & 0 \\ & 1 & 2 & 0 \\ & & 4 & 0 \\ \text{Sym} & & & 0 \end{array} \right] \end{array} \begin{array}{c} \hat{\varphi}_{i=1} \\ \hat{\varphi}_{i=2} \\ \hat{\varphi}_{i=3} \\ \varphi_{i\geq 4} \end{array} \qquad [9.26]$$

It is worth remembering here that the value of penalty parameters representing artificial springs or artificial masses are usually defined with positive values. But negative values of both types of penalty parameters can also be used to model constraints [ILA 99] and [ILA 05]. The effect of using negative penalty parameters is that the convergence changes from lower bounds to upper bounds when using artificial springs, and form upper bounds to lower bounds when using artificial masses.

9.5. Modes of vibration

The solution of the eigenproblem defined in equation [9.12b] gives the eigenvalues of the system and their corresponding eigenvectors. The natural frequencies are the square root of the eigenvalues, while the eigenvectors define the vibration mode shape. When using the RRM, the vibration modes of the continuum (whole structural element) are calculated using the eigenvectors. Thus, to calculate a modal shape of vibration of the beam as a function of the coordinate it is necessary to carry out the summation of the set of admissible functions evaluated at the point of interest times their corresponding term of the eigenvector as shown in the following equation:

$$Mode_i(\xi) = \sum_{j=1}^{n} X_{i,j}\varphi_j(\xi) = X_{i,j} \bullet \varphi_j(\xi) \qquad [9.27]$$

The summation can also be considered a dot product as shown on the right hand side of the equation [9.27].

9.6. Results

The results of beams with classical boundary conditions (F, S and C) as well as sliding conditions (G) are presented in comparison with the values given by Karnovsky and Lebed [KAR 00]. To identify the various support cases two letters are used representing the boundary conditions at the two ends of the beam. For instance, FF indicates free–free beam. Convergence of the results by increasing the number of terms or increasing the value of the penalty parameters is demonstrated in the next section.

9.6.1. *Free–free beam*

The first set of results presented here are for a free–free beam increasing the number of terms in the series. Because the series has a constant term, a linear term, a quadratic term and a series of cosine functions, it is necessary to include at least four terms in the series to include the first cosine function of the series. As the number of terms increases, the number of cosine functions included in the admissible function increases. For the free–free beam the stiffness matrix has non-zero values only on the main diagonal and the mass matrix has non-zero off-diagonal terms only in the first three rows and columns which correspond to the polynomial terms.

Table 9.5 shows the results of a free–free beam. The results using the proposed set of admissible functions give two zero frequency parameters (corresponding to a rigid body translation and a rigid body rotation) while, as expected, the last two frequency parameters do not converge to exact results for any kind of number of terms in the series. The rest of the results converge to non-rigid body natural frequencies of the beam. The results presented in the last row are given by

Karnovsky and Lebed [KAR 00]. Table 9.5 shows how the accuracy of the solution increases as the number of terms increases. For example, the first two non-rigid body frequency parameters obtained using 6 terms in the series match the results given in [KAR 00] only up to the third significant number, while results using 50 terms matched the results given in [KAR 00].

λ_i	3	4	5	6	7
Number of terms					
6	4.73074149	7.85778198	13.1006371	17.2089566	NA
7	4.73006967	7.85778198	11.0079022	17.2089566	21.3523843
8	4.73006967	7.85352551	11.0079022	14.1600886	21.3523843
9	4.73004390	7.85352551	10.9968442	14.1600886	17.3140978
10	4.73004390	7.85325117	10.9968442	14.1401604	17.3140978
50	4.73004074	7.85320462	10.9956078	14.1371655	17.2787597
200	4.73004074	7.85320462	10.9956078	14.1371655	17.2787597
[KAR 00]	4.73004074	7.85320462	10.9956078	14.1371655	17.2787597

Table 9.5. *Frequency parameters λ_i of a FF beam with increasing number of terms in the series*

9.6.2. *Clamped–clamped beam using 250 terms*

Convergence due to the value of the penalty parameter was investigated comparing the results for a CC beam using 250 terms with different penalty values. Results presented in this work correspond to two types of penalty parameters: (1) negative stiffnesses presented in Table 9.6 and (2) positive masses presented in Table 9.7 In both cases calculations for penalty parameters with several values of the series 10^p were computed, where $p \in [0,1,2,3,...]$. *Both cases give upper bound solutions.*

Results in Table 9.6 show that the first frequency is the first mode to converge when penalty parameters of stiffness type are used and that as the absolute magnitude of the penalty parameter increases, better approximations of the frequency parameters were obtained. Checking for solutions free of ill-conditioning (following monotonic convergence), it was observed that the best approximation of the upper bound frequency parameters was found using stiffness with a value of -10^9. Beyond this value, the calculated natural frequency does not continue to decrease with increasing negative stiffness and this can only be explained by round-off errors associated with large penalty terms. Results with even higher magnitude of stiffness show that the results do not change drastically such as it would be expected in ill-conditioning until a very high penalty parameter value is used (-10^{13} or higher) when the results for the second frequency began to deviate significantly from the values given in [KAR 00]. This shows that there is a limitation in the maximum penalty parameter value that can be used due to numerical problems. This applies to all problems where penalty parameters are used. Results in square brackets in Tables 9.6 and 9.7 are the benchmark results used for comparison and these are given in [KAR 00].

$\hat{k}$	Mode number					
	1	2	3	4	5	6
-10^3	4.9488	8.6122	12.1548	15.4912	18.7299	21.9248
-10^4	4.7515	7.9506	11.2456	14.6065	17.9869	21.3396
-10^5	4.7322	7.8631	11.0223	14.1930	17.3788	20.5808
-10^6	4.7303	7.8542	10.9983	14.1428	17.2891	20.4373
-10^7	4.7301	7.8533	10.9959	14.1377	17.2798	20.4221
-10^8	4.7300	7.8532	10.9956	14.1372	17.2789	20.4205
-10^9	4.7300	7.8532	10.9956	14.1372	17.2788	20.4204
-10^{10}	4.7300	7.8539	10.9956	14.1370	17.2788	20.4203
-10^{11}	4.7300	7.8637	10.9956	14.1378	17.2788	20.4207
-10^{12}	4.7301	7.8554	10.9955	14.1407	17.2788	20.4232
-10^{13}	4.7303	8.3046	10.9959	14.4403	17.2792	20.4234
[KAR 00]	4.7300	7.8532	10.9956	14.1372	17.2788	20.4204

Table 9.6. *Frequency parameters of a CC beam using 250 terms and stiffness penalty parameters*

Table 9.7 shows that when using small inertial penalty parameters in beams, all the frequencies give a good approximation of the exact results. For the CC beam case all frequencies were obtained within 1% of the exact value using an inertial penalty parameter equal to 10, which means just applying 10 times the mass of the beam at a point as an inertial penalty parameter. Although the best approximation that is free of ill-conditioning was obtained with an inertial penalty parameter value of 10^7, the results remained unchanged up to the fourth decimal included in Table 9.7 for the next penalty parameters (up to 10^{10}). Then the results started to change when the inertial penalty parameter reached the value 10^{11}, giving different values but only on the fourth decimal. The solution started to deviate significantly from the values given in [KAR 00] when the penalty parameter was assigned a value of 10^{14}. In all beam cases using inertial penalty parameters, the results included two extra rigid-body frequencies (zero frequencies), because there is no restraint attached to the rigid-bodies. These extra zero frequencies are not included in the tables presented in this work.

Comparison of results of Tables 9.6 and 9.7 suggests that the frequency results are obtained with lower magnitude of the penalty values using artificial inertia, rather than artificial stiffness. Furthermore, negative artificial stiffness has critical values that give extra zero frequencies as explained in earlier chapters. For these reasons, the inertial penalty parameters will be used in the rest of this chapter.

$\hat{m}$	Mode number					
	1	2	3	4	5	6
10	4.7715	7.8787	11.0138	14.1513	17.2903	20.4301
10^2	4.7343	7.8558	10.9974	14.1386	17.2799	20.4213
10^3	4.7305	7.8535	10.9958	14.1373	17.2789	20.4205
10^4	4.7301	7.8532	10.9956	14.1372	17.2788	20.4204
10^5	4.7300	7.8532	10.9956	14.1372	17.2788	20.4204
10^6	4.7300	7.8532	10.9956	14.1372	17.2788	20.4204
10^7	4.7300	7.8532	10.9956	14.1372	17.2788	20.4204
10^8	4.7300	7.8532	10.9956	14.1372	17.2788	20.4204
10^9	4.7300	7.8532	10.9956	14.1372	17.2788	20.4204
10^{10}	4.7300	7.8532	10.9956	14.1372	17.2788	20.4204
10^{11}	4.7300	7.8533	10.9956	14.1370	17.2787	20.4202
10^{12}	4.7300	7.8522	10.9959	14.1386	17.2791	20.4210
10^{13}	4.7364	7.8614	11.0001	14.1565	17.2844	20.4149
10^{14}	4.7300	7.9935	10.9552	14.3009	17.1401	20.4839
[KAR 00]	4.7300	7.8532	10.9956	14.1372	17.2788	20.4204

Table 9.7. *Frequency parameters of a CC beam using 250 terms and inertial penalty parameters*

9.6.3. *Beam with classical and sliding boundary conditions using inertial restraints to model constraints at the edges of the beam*

Table 9.8 contains the best approximations to the non-dimensional frequency parameters of beams with all combinations of classical boundary conditions using 10, 50, 250 and 1,000 terms and positive inertial penalty parameters with values 10^p where $p \in [0,1,2,3,...]$. Table 9.9 presents results for cases that contain sliding conditions (denoted by G), which may be useful when symmetry is applied to a structure; sliding conditions allow lateral displacement while rotation is fixed. Results in Tables 9.8 and 9.9 are compared to the results given in [KAR 00]. Results using at least 250 terms in Tables 9.8 and 9.9 are the same as those published in [KAR 00], with one exception: the second frequency of the CG case using the present method is slightly lower. The published figure in [KAR 00] is 5.4988. However, by symmetry one can deduce the CG case results from that of the CC case, where each frequency parameter for the CG case has a frequency parameter of twice the magnitude corresponding to the symmetrical modes of CC beam. Using the second symmetrical frequency parameter for the CC beam, it can be seen that the second frequency parameter for the CG beam should be 5.4978, which is the result obtained using the present method. Therefore, we have given the corrected figure in Table 9.8. With this correction, there is complete agreement between the results in [KAR 00] and the present results.

For most beam cases, penalty parameters that gave the best approximations to the frequency parameters do not change as the number of terms included in the solution increased. In most cases solutions converge up to the fourth decimal place, well before the monotonic convergence is interrupted. For instance, for GG beams using positive inertial penalty parameters, a value of 10^3 is enough to get all sets of results presented in Table 9.9. Furthermore, as mentioned for the CC beam, in beam problems, it was found that all frequencies gave good approximations to the exact result using very small inertial penalty parameter values. For instance with an inertial penalty parameter of 10 (ten times that of the beam itself) for all beam cases included in Tables 9.8 and 9.9 and for any number of terms used in this work, the first frequency has a deviation lower than 3% in comparison with the exact solution. It should be emphasized here that this deviation corresponds to the first frequency, which is the last one to converge, and is therefore the worst case scenario. The results for all other frequencies present a lower deviation.

Case	$\hat{m}$	Mode number					
		1	2	3	4	5	6
SF							
10	10^6	0	3.9271	7.0716	10.2222	13.3774	16.5883
50	10^6	0	3.9266	7.0686	10.2102	13.3518	16.4935
250	10^6	0	3.9266	7.0686	10.2102	13.3518	16.4934
1,000	10^6	0	3.9266	7.0686	10.2102	13.3518	16.4934
[KAR 00]		0	3.9266	7.0686	10.2102	13.3518	16.4934
CF							
10	10^6	1.8752	4.6958	7.8623	11.0192	14.1795	17.4005
50	10^6	1.8751	4.6941	7.8548	10.9956	14.1373	17.2791
250	10^6	1.8751	4.6941	7.8548	10.9955	14.1372	17.2788
1,000	10^6	1.8751	4.6941	7.8548	10.9955	14.1372	17.2788
[KAR 00]		1.8751	4.6941	7.8548	10.9955	14.1372	17.2788
SS							
10	10^5	3.1422	6.2866	9.4447	12.6010	15.8510	19.0328
50	10^5	3.1416	6.2832	9.4248	12.5665	15.7082	18.8500
250	10^5	3.1416	6.2832	9.4248	12.5664	15.7080	18.8496
1,000	10^5	3.1416	6.2832	9.4248	12.5664	15.7080	18.8496
[KAR 00]		3.1416	6.2832	9.4248	12.5664	15.7080	18.8496
SC							
10	10^6	3.9283	7.0757	10.2438	13.4047	16.6759	19.8580
50	10^6	3.9266	7.0686	10.2103	13.3520	16.4938	19.6357
250	10^6	3.9266	7.0686	10.2102	13.3518	16.4934	19.6350
1,000	10^6	3.9266	7.0686	10.2102	13.3518	16.4934	19.6350
[KAR 00]		3.9266	7.0686	10.2102	13.3518	16.4934	19.6350
CC							
10	10^7	4.7340	7.8658	11.0472	14.2118	17.4959	20.6671
50	10^7	4.7301	7.8533	10.9958	14.1375	17.2794	20.4214
250	10^7	4.7300	7.8532	10.9956	14.1372	17.2788	20.4204
1,000	10^7	4.7300	7.8532	10.9956	14.1372	17.2788	20.4204
[KAR 00]		4.7300	7.8532	10.9956	14.1372	17.2788	20.4204

Table 9.8. *Frequency parameters of beams with classical boundary conditions*

Case	$\hat{m}$	Mode number					
		1	2	3	4	5	6
FG							
10	10^4	0	2.3650	5.4978	8.6395	11.7817	14.9265
50	10^4	0	2.3650	5.4978	8.6394	11.7810	14.9226
250	10^4	0	2.3650	5.4978	8.6394	11.7810	14.9226
1,000	10^4	0	2.3650	5.4978	8.6394	11.7810	14.9226
[KAR 00]		0	2.3650	5.4978	8.6394	11.7810	14.9226
GG							
10	10^{16}	0	3.1416	6.2832	9.4248	12.5664	15.7080
50	10^9	0	3.1416	6.2832	9.4248	12.5664	15.7080
250	10^7	0	3.1416	6.2832	9.4248	12.5664	15.7080
1,000	10^8	0	3.1416	6.2832	9.4248	12.5664	15.7080
[KAR 00]		0	3.1416	6.2832	9.4248	12.5664	15.7080
SG							
10	10^6	1.5708	4.7133	7.8584	11.0095	14.1728	17.3639
50	10^6	1.5708	4.7124	7.8540	10.9956	14.1373	17.2789
250	10^6	1.5708	4.7124	7.8540	10.9956	14.1372	17.2788
1,000	10^6	1.5708	4.7124	7.8540	10.9956	14.1372	17.2788
[KAR 00]		1.5708	4.7124	7.8540	10.9956	14.1372	17.2788
CG							
10	10^7	2.3652	5.5005	8.6498	11.8074	14.9774	18.1666
50	10^7	2.3650	5.4978	8.6394	11.7811	14.9228	18.0645
250	10^7	2.3650	5.4978	8.6394	11.7810	14.9226	18.0642
1,000	10^7	2.3650	5.4978	8.6394	11.7810	14.9226	18.0642
[KAR 00]		2.3650	5.4978	8.6394	11.7810	14.9226	18.0642

Table 9.9. *Frequency parameters of beams with classical and sliding boundary conditions*

The results for GG beams shown in Table 9.9 converged to the values given in [KAR 00] using non-dimensional masses with a magnitude of 10^8, but monotonic convergence was interrupted only with values higher than 10^{16}. In all cases, the results included two extra rigid body frequencies (zero frequencies), because the constraints are modeled with artificial masses and the stiffness matrix corresponds to the stiffness of an unconstraint beam, in which first two rows and columns are full of zeroes. The two extra rigid-body mode results are not included in any set of results.

For all beam cases containing constraints and 10 to 250 terms, inertial penalty parameters with a value of 105 gave very good approximations of the frequency parameters (error smaller than 1.5%, if the last two frequency parameters in the solution are not taken into account).

9.7. Modes of vibration

Figure 9.2 shows the first seven modes of vibration of an unconstrained (free–free) beam. These results were obtained by setting the number of admissible functions to 13 and evaluating the modes at 200 points to obtain smooth curves.

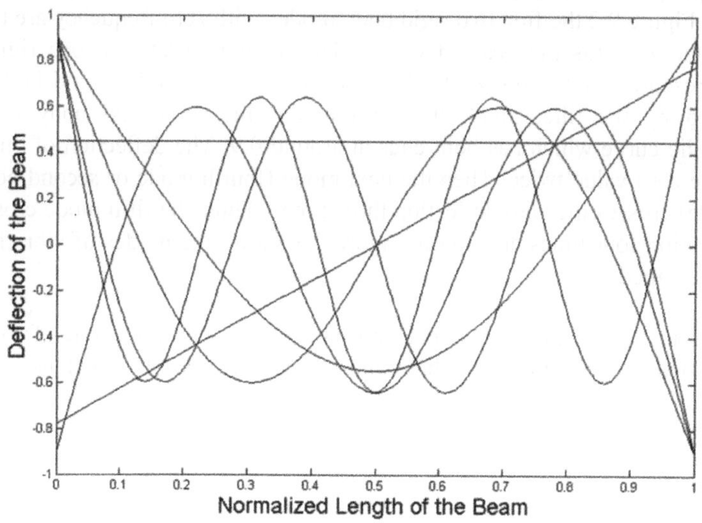

Figure 9.2. *Modes of vibration of a free–free beam*

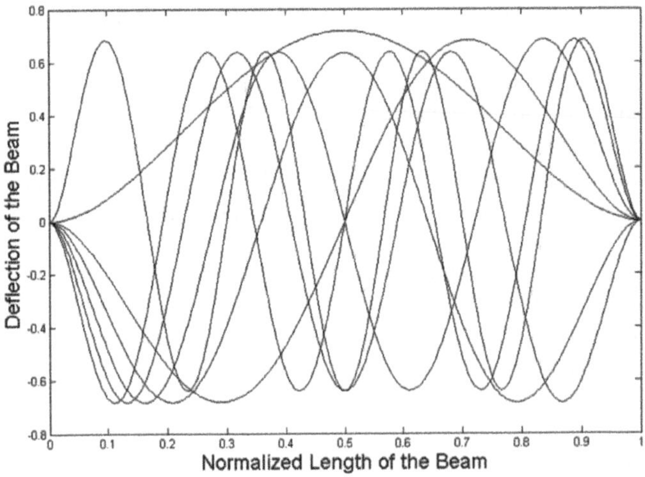

Figure 9.3. *Modes of vibration of a clamped-clamped beam*

A simple procedure can be used to identify the corresponding frequencies of the modes. In Figure 9.2 the first two rigid body modes with zero frequency are the only two straight lines (as the beam does not deform, it translates in one rigid body mode represented by the horizontal line and rotates in the other rigid body mode represented by the tilted line); the third mode (first non-rigid body mode) is the parabolic curve which has both ends at around 0.9. The deflection of this mode crosses the zero value twice. Then the next mode (fourth mode or second non-rigid body mode) crosses the zero deflection three times. Thus, the fifth mode crosses the zero deflection four times and so on. Figure 9.3 shows the modes of vibration of a clamped-clamped beam.

Results in this chapter show that very accurate results for natural frequencies were obtained conveniently using the proposed set of admissible functions.

10

Natural Frequencies and Modes of Plates of Rectangular Planform

10.1. Introduction

In this chapter we will see how to apply the Rayleigh–Ritz procedure to solve vibration problems of thin plates that follow Kirchhoff's hypotheses [VEN 01] using the same set of admissible functions used for beams in the previous chapter. There are 55 different combinations of the four boundary conditions, namely, unconstrained or free (F), simply supported (S), clamped (C) and guided (G). As some features of the procedure have been discussed in the previous two chapters, and the number of sets of boundary conditions is large, we will present only the essential parts of the procedure and the results, including the modes, to enable readers to repeat implementing this procedure or developing it for other complex problems.

10.2. Theoretical derivations of the eigenvalue problems

Consider a rectangular plate as shown in Figure 10.1 with width a, height b, constant thickness h and flexural rigidity D defined as:

$$D = \frac{Eh^3}{12\left(1-\nu^2\right)}$$

where ν is Poisson's ratio and E is Young's modulus.

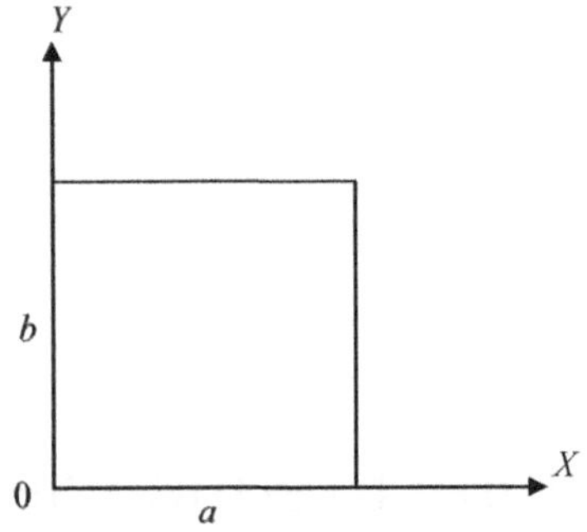

Figure 10.1. *Completely free rectangular plate*

The set of admissible functions for beams presented in the previous chapters are used to model the deflection of the plate. In this case, one admissible function is needed for each direction. Then the two sets of admissible functions for plates used in this work are:

– set of admissible functions in the x direction

$$\varphi_i(x) = 1 \quad \text{if} \quad i = 1 \qquad [10.1a]$$

$$\varphi_i(x) = \frac{x}{L} \quad \text{if} \quad i = 2 \qquad [10.1b]$$

$$\varphi_i(x) = \left(\frac{x}{L}\right)^2 \quad \text{if} \quad i = 3 \qquad [10.1c]$$

$$\varphi_i(x) = \cos\frac{(i-3)\pi x}{a} \quad \text{if} \quad i = 4, 5, 6, \ldots, n \qquad [10.1d]$$

– set of admissible functions in the y direction

$$\psi_j(y) = 1 \quad \text{if} \quad j = 1 \qquad [10.1e]$$

$$\psi_j(y) = \frac{y}{L} \quad \text{if} \quad j = 2 \qquad [10.1f]$$

$$\psi_j(y) = \left(\frac{y}{L}\right)^2 \quad \text{if} \quad j = 3 \qquad [10.1g]$$

$$\psi_j(y) = \cos\frac{(j-3)\pi y}{b} \quad \text{if} \quad j = 4,5,6,\ldots,n \qquad [10.1h]$$

It is also convenient to present the sets of admissible functions in non-dimensional form as follows:

– non-dimensional set of admissible functions in the x direction

$$\hat{\varphi}_i(\xi) = 1 \quad \text{if} \quad i = 1 \qquad [10.2a]$$

$$\hat{\varphi}_i(\xi) = \xi \quad \text{if} \quad i = 2 \qquad [10.2b]$$

$$\hat{\varphi}_i(\xi) = \xi^2 \quad \text{if} \quad i = 3 \qquad [10.2c]$$

$$\hat{\varphi}_i(\xi) = \cos(i-3)\pi\xi \quad \text{if} \quad i = 4,5,6,\ldots,n \qquad [10.2d]$$

where: $\xi = x/a$, and

– non-dimensional set of admissible functions in the y direction

$$\hat{\psi}_j(y) = 1 \quad \text{if} \quad j = 1 \qquad [10.2e]$$

$$\hat{\psi}_j(y) = \eta \quad \text{if} \quad j = 2 \qquad [10.2f]$$

$$\hat{\psi}_j(y) = \eta^2 \quad \text{if} \quad j = 3 \qquad [10.2g]$$

$$\psi_j(y) = \cos(j-3)\pi\eta \quad \text{if} \quad j = 4,5,6,\ldots,n \qquad [10.2h]$$

with $\eta = y/b$.

Thus, the deflected shape of the plate $W(x,y)$ is

$$W(x,y) = \sum_{i=1}^{n}\sum_{k=1}^{m} a_{ik}\varphi_i(x)\psi_k(y), \qquad [10.3a]$$

where a_{ik} are unknown coefficients, while n and m are the number of admissible functions used in the x and y directions, respectively.

Therefore, the non-dimensional deflected shape of the plate $\hat{W}(\xi,\eta)$ is:

$$\hat{W}(\xi,\eta)=\sum_{i=1}^{n}\sum_{k=1}^{m}a_{ik}\hat{\varphi}_i(\xi)\hat{\psi}_k(\eta), \qquad [10.3b]$$

When free vibrations are assumed, the motion of the plate is defined [LEI 73] as:

$$w=W\cos(\omega t) \qquad [10.4]$$

Once the deflection shape of the plate is modeled the energy terms can be computed. The maximum kinetic energy $T_{\max}$ of a plate is defined as:

$$T_{\max}=\frac{\rho h\omega^2}{2}\int_0^a\int_0^b W^2dxdy, \qquad [10.5a]$$

where ρ is the mass density of the material of the plate. The non-dimensional maximum kinetic energy is obtained dividing $T_{\max}$ by ρh and expressing $W(x,y)$ in non-dimensional coordinates as defined in equation [10.3b]:

$$\hat{T}_{\max}=\int_0^1\int_0^1\frac{\omega^2}{2}\hat{W}^2d\xi d\eta, \qquad [10.5b]$$

The non-dimensional kinetic energy function $\hat{\Gamma}_{\max}$ is:

$$\hat{\Gamma}_{\max}=\frac{\hat{T}_{\max}}{\omega^2}=\int_0^1\int_0^1\frac{\hat{W}^2}{2}d\xi d\eta, \qquad [10.5c]$$

The strain energy of bending and twisting $V_{\max}$ is defined as:

$$V_{\max}=\frac{1}{2}D\int_0^a\int_0^b\left[\left(\frac{\partial^2W}{\partial x^2}+\frac{\partial^2W}{\partial x^2}\right)^2-2(1-\nu)\left(\frac{\partial^2W}{\partial x^2}\frac{\partial^2W}{\partial y^2}-\left(\frac{\partial^2W}{\partial y\partial x}\right)^2\right)\right]dxdy, \qquad [10.6a]$$

The non-dimensional strain energy of bending and twisting $\hat{V}_{\max}$ is obtained dividing $V_{\max}$ by D and expressing $W(x,y)$ in non-dimensional coordinates:

$$\hat{V}_{\max}=\frac{1}{2}\int_0^1\int_0^1\left[\left(\frac{\partial^2\hat{W}}{\partial\xi^2}+\alpha^2\frac{\partial^2\hat{W}}{\partial\eta^2}\right)^2-2(1-\nu)\alpha^2\left(\frac{\partial^2\hat{W}}{\partial\xi^2}\frac{\partial^2\hat{W}}{\partial\eta^2}-\left(\frac{\partial^2\hat{W}}{\partial\xi\partial\eta}\right)^2\right)\right]d\xi d\eta \qquad [10.6b]$$

where the aspect ratio of the plate is defined as $\alpha = a/b$. In this work only results for square plates are presented, which means that for all cases presented here $\alpha = 1$. Also, for all solutions presented in this chapter $\nu = 0.3$. The non-dimensional mass matrix is obtained minimizing the maximum kinetic function given in equation [10.5c], while the non-dimensional stiffness matrix is obtained by minimizing the strain energy given in equation [10.6b], respectively.

Thus the terms of the non-dimensional mass matrix and the stiffness matrix are built by the terms of fourth order arrays

$$\hat{M}_{ijkl} = E_{mi}^{(0,0)} F_{nj}^{(0,0)}, \quad \text{and} \tag{10.7}$$

$$\begin{aligned} \hat{K}_{ijkl} &= E_{mi}^{(2,2)} F_{nj}^{(0,0)} + \alpha^4 E_{mi}^{(0,0)} F_{nj}^{(2,2)} + \\ &\nu\alpha^2 \left[E_{mi}^{(0,2)} F_{nj}^{(2,0)} + E_{mi}^{(2,0)} F_{nj}^{(0,2)} \right] + 2(1-\nu)\alpha^2 E_{mi}^{(1,1)} F_{nj}^{(1,1)}, \end{aligned} \tag{10.8}$$

where

$$E_{mi}^{(r,s)} = \int_0^1 \left(\frac{d^r \hat{\varphi}_n}{d\xi^m} \right) \left(\frac{d^s \hat{\varphi}_j}{d\xi^i} \right) d\xi \text{ and } \qquad F_{nj}^{(r,s)} = \int_0^1 \left(\frac{d^r \hat{\psi}_n}{d\eta^r} \right) \left(\frac{d^s \hat{\psi}_j}{d\eta^s} \right) d\eta$$

Noting that $\hat{W}(\xi,\eta)$ is a linear combination of terms, the non-dimensional mass matrix $\hat{\mathbf{M}}_{rs}$ and stiffness matrix $\hat{\mathbf{K}}_{rs}$ will be obtained combining every term in the summation with subscripts ij with a term with subscripts kl, using the relationships $r = j + (i-1)j_{\max}$ and $s = l + (k-1)l_{\max}$. $\hat{\mathbf{M}}_{rs}$ and $\hat{\mathbf{K}}_{rs}$ are square matrices.

The frequency parameters can be obtained by substituting the mass and stiffness in the following equation (omitting the subscripts r and s):

$$\hat{\mathbf{K}} - \lambda^2 \hat{\mathbf{M}} = 0 \tag{10.9}$$

Results in this work are presented using the non-dimensionalized frequency parameter λ, which is defined as

$$\lambda = \frac{\omega\sqrt{\rho h a^4}}{D} \tag{10.10}$$

An advantage of the procedure described in this chapter is that the same set of admissible functions is used in both directions and if the same number of terms are used in both directions:

$$E_{mi}^{(r,s)} = F_{nj}^{(r,s)} \tag{10.11}$$

This reduces the amount of work to calculate the mass matrix $\hat{\mathbf{M}}_{rs}$ and stiffness matrix $\hat{\mathbf{K}}_{rs}$. Furthermore, from equations [10.7] and [10.8] it is clear that to compute the mass matrix $\hat{\mathbf{M}}_{rs}$ and stiffness matrices $\hat{\mathbf{K}}_{rs}$ it is necessary to compute $E_{mi}^{(0,0)}$, $E_{mi}^{(2,2)}$, $E_{mi}^{(0,2)}$, $E_{mi}^{(2,0)}$ and $E_{mi}^{(1,1)}$. In the beam chapter $E_{mi}^{(0,0)}$, $E_{mi}^{(2,2)}$ were computed and $E_{mi}^{(2,0)} = E_{mi}^{(0,2)\mathrm{T}}$. Thus, having solved vibration problems of beams only $E_{mi}^{(2,0)}$ and $E_{mi}^{(1,1)}$ need to be computed to solve vibration problems of plates. As before only the first four rows and columns are enough to define $E_{mi}^{(2,0)}$ and $E_{mi}^{(1,1)}$ in parametric form.

A procedure to obtain the terms $E_{mi}^{(2,0)}$:

$$E_{mi}^{(2,0)} = \int_0^1 \left(\frac{d^2\hat{\varphi}_n}{d\xi^m}\right)\left(\frac{d^0\hat{\varphi}_j}{d\xi^i}\right) d\xi = \int_0^1 \left(\frac{d^2\hat{\varphi}_n}{d\xi^m}\right)\hat{\varphi}_j d\xi$$

starts by gathering the products of the second derivatives $\hat{\varphi}_m''$ and functions $\hat{\varphi}_i$ in Table 10.1 shown below:

$m \setminus i$	$\hat{\varphi}_{i=1}$	$\hat{\varphi}_{i=2}$	$\hat{\varphi}_{i=3}$	$\hat{\varphi}_{i\geq 4}$
$\hat{\varphi}_{m=1}''$	0	0	0	0
$\hat{\varphi}_{m=2}''$	0	0	0	0
$\hat{\varphi}_{m=3}''$	2	$2x$	$2x^2$	$2\cos((i-3)\pi\xi)$
$\hat{\varphi}_{m\geq 4}''$	$-(m-3)^2\pi^2$ $\cos((m-3)\pi\xi)$	$-(m-3)^2\pi^2 x$ $\cos((m-3)\pi\xi)$	$-(m-3)^2\pi^2 x^2$ $\cos((m-3)\pi\xi)$	$-(m-3)^2\pi^2$ $\cos((m-3)\pi\xi)$ $\cos((i-3)\pi\xi)$

Table 10.1. *Products of the second derivatives of the set of admissible functions $\hat{\varphi}_m''$ and the set of admissible functions $\hat{\varphi}_i$ to build matrix $E_{mi}^{(2,0)}$*

$m \backslash i$	$\hat{\varphi}_{i=1}$	$\hat{\varphi}_{i=2}$	$\hat{\varphi}_{i=3}$	$\hat{\varphi}_{i\geq 4}$
$\hat{\varphi}''_{m=1}$	0	0	0	0
$\hat{\varphi}''_{m=2}$	0	0	0	0
$\hat{\varphi}''_{m=3}$	2	1	2/3	0
$\hat{\varphi}''_{m\geq 4}$	0	$1-(\cos((i-3)\pi))$	$-2\cos((i-3)\pi)$	$-\frac{1}{2}(m-3)^2\pi^2$

Table 10.2. $E_{mi}^{(2,0)}$ *terms*

Carrying out the integrals of the functions in Table 10.1 with respect to ξ, and using limits of integration 0 and 1, gives the terms shown in Table 10.2. All integrals are very easy to carry out by hand and are very similar to integrals carried out in the previous chapter. The integrals of terms $E_{31}^{(2,0)}$, $E_{32}^{(2,0)}$ and $E_{33}^{(2,0)}$ are simple polynomials; the integrals of $E_{34}^{(2,0)}$ and $E_{41}^{(2,0)}$ give each a sine function. Thus $E_{34}^{(2,0)}$ and $E_{41}^{(2,0)}$ are zero because sine functions are zero at both limits of integration (0 and 1); integrals for the last three terms $E_{42}^{(2,0)}$, $E_{43}^{(2,0)}$ and $E_{44}^{(2,0)}$ are similar to the integrals carried out in the previous chapter for terms $M_{24}^{(0,0)}$, $M_{34}^{(0,0)}$ and $M_{44}^{(0,0)}$. As before, $M_{m\geq 4,i\geq 4}^{(0,0)}$ has non-zero values only along the main diagonal ($i = j$).

It is worth noting that although $E_{mi}^{(2,0)}$ is not symmetric, the stiffness matrix is symmetric. Similarly a procedure to obtain the terms $E_{mi}^{(1,1)}$:

$$E_{mi}^{(1,1)} = \int_0^1 \left(\frac{d\hat{\varphi}_n}{d\xi^m} \right) \left(\frac{d\hat{\varphi}_j}{d\xi^i} \right) d\xi$$

starts by gathering the products of the derivatives $\hat{\varphi}'_m$ and $\hat{\varphi}'_i$ as shown in Table 10.3.

Carrying out the integrals of the functions in Table 10.3 with respect to ξ, and using limits 0 and 1, gives the terms shown in Table 10.4. All integrals are very easy to carry out by hand and very similar to integrals carried out in the previous chapter. It is worth noting once again that values for $E_{m\geq 4,i\geq 4}^{(1,1)}$ are non-zero only if $i = j$.

$m \backslash i$	$\hat{\varphi}'_{i=1}$	$\hat{\varphi}'_{i=2}$	$\hat{\varphi}'_{i=3}$	$\hat{\varphi}'_{i\geq 4}$
$\hat{\varphi}'_{m=1}$	0	0	0	0
$\hat{\varphi}'_{m=2}$	0	1	$2x$	$-(i-3)\pi$ $\sin((i-3)\pi\xi)$
$\hat{\varphi}'_{m=3}$	0	$2x$	$4x^2$	$-2(i-3)\pi x$ $\sin((i-3)\pi\xi)$
$\hat{\varphi}'_{m\geq 4}$	0	$-(m-3)\pi$ $\sin((m-3)\pi\xi)$	$-2(m-3)\pi x$ $\sin((m-3)\pi\xi)$	$-4(m-3)(i-3)\pi^2 x^2$ $\sin((m-3)\pi\xi)$ $\sin((i-3)\pi\xi)$

Table 10.3. *Products of the first derivatives of the set of admissible functions $\hat{\varphi}'_m$ and the set of admissible functions $\hat{\varphi}'_i$ to build matrix $E_{mi}^{(1,1)}$*

$m \backslash i$	$\hat{\varphi}'_{i=1}$	$\hat{\varphi}'_{i=2}$	$\hat{\varphi}'_{i=3}$	$\hat{\varphi}'_{i\geq 4}$
$\hat{\varphi}'_{m=1}$	0	0	0	0
$\hat{\varphi}'_{m=2}$	0	1	1	$(\cos((i-3)\pi))-1$
$\hat{\varphi}'_{m=3}$	0	1	$4/3$	$2\cos((i-3)\pi)$
$\hat{\varphi}'_{m\geq 4}$	0	$(\cos((m-3)\pi))-1$	$2\cos((m-3)\pi)$	$\frac{1}{2}(m-3)^2\pi^2$

Table 10.4. $E_{mi}^{(1,1)}$ *terms*

10.3. Derivation of the eigenvalue problem for plates containing classical constraints along its edges

Simply supported (S), clamped (C) and guided (G) constraints can be modeled by restraints represented by a series of translational and/or rotational springs or alternatively by masses and/or moments of inertia. Translational springs or masses are used to model simply supported (S) boundary conditions; translational and rotational springs or masses and moment of inertia are used to model clamped (C) boundary conditions, while rotational constraints (rotational springs or moments of inertia) are used to model guided (G) constraints.

The number of masses or springs per edge is equal to the number of terms used in each direction. Considering that all springs have the same non-dimensional stiffness coefficient defined as:

$$\hat{p} = \frac{k_i a^3}{D} = \frac{kr_i a}{D} \qquad [10.12]$$

The additional strain energy in non-dimensional form of the series of translational springs at $\xi = 0$, $\xi = 1$, $\eta = 0$ and $\eta = 1$ is

$$\hat{V}_{translational\_springs} = \frac{1}{2}\int_0^1 \left(\hat{p}\hat{W}^2\Big|_{\xi=0} + \hat{p}\hat{W}^2\Big|_{\xi=1} \right) d\eta + \frac{1}{2}\int_0^1 \left(\hat{p}\hat{W}^2\Big|_{\eta=0} + \hat{p}\hat{W}^2\Big|_{\eta=1} \right) d\xi \qquad [10.13]$$

and the additional strain energy of rotational springs at $\xi = 0$, $\xi = 1$, $\eta = 0$ and $\eta = 1$ is:

$$\begin{aligned} \hat{V}_{rotational\_springs} = {} & \frac{1}{2}\int_0^1 \left(\hat{p}\left(\frac{\partial \hat{W}}{\partial \xi}\right)^2\Bigg|_{\xi=0} + \hat{p}\left(\frac{\partial \hat{W}}{\partial \xi}\right)^2\Bigg|_{\xi=1} \right) d\eta + \\ & \frac{1}{2}\int_0^1 \left(\hat{p}\left(\frac{\partial \hat{W}}{\partial \eta}\right)^2\Bigg|_{\eta=0} + \hat{p}\left(\frac{\partial \hat{W}}{\partial \eta}\right)^2\Bigg|_{\eta=1} \right) d\xi \end{aligned} \qquad [10.14]$$

If masses are used, the terms added to the maximum kinetic energy are the same as the terms above, but the translational and rotational springs should be substituted by masses and rotational inertia and the non-dimensional penalty parameter is equal to:

$$\hat{p} = \frac{m_i}{\rho hab} = \frac{I_i}{\rho h a^2 b^2} \qquad [10.15]$$

To model a constraint at a point (ξ_1, η_1) the values for ξ_1 and η_1 should be substituted in the expression for $\hat{W}$ in the last two equations above, respectively.

Defining $\hat{P}_{edge,ijkl}$ as the fourth dimensional array with all penalty terms obtained by adding all terms resulting from the minimization of equations [10.13] and [10.4] gives

$$\hat{P}_{edge,ijkl} = \hat{p}\begin{bmatrix} \hat{\varphi}_m(0)\hat{\varphi}_i(0)F_{nj}^{(0,0)} + \hat{\varphi}_m(1)\hat{\varphi}_i(1)F_{nj}^{(0,0)} + \\ E_{mi}^{(0,0)}\hat{\psi}_n(0)\hat{\psi}_j(0) + E_{mi}^{(0,0)}\hat{\psi}_n(1)\hat{\psi}_j(1) + \\ \dfrac{\partial\hat{\varphi}_m(0)}{\partial\xi}\dfrac{\partial\hat{\varphi}_i(0)}{\partial\xi}F_{nj}^{(0,0)} + \dfrac{\partial\hat{\varphi}_m(1)}{\partial\xi}\dfrac{\partial\hat{\varphi}_i(1)}{\partial\xi}F_{nj}^{(0,0)} + \\ E_{mi}^{(0,0)}\dfrac{\partial\hat{\psi}_n(0)}{\partial\eta}\dfrac{\partial\hat{\psi}_j(0)}{\partial\eta} + E_{mi}^{(0,0)}\dfrac{\partial\hat{\psi}_n(0)}{\partial\eta}\dfrac{\partial\hat{\psi}_j(0)}{\partial\eta} \end{bmatrix} \qquad [10.16]$$

The terms $E_{mi}^{(0,0)} = F_{nj}^{(0,0)}$ were already obtained to solve beam problems and it should be obvious to the reader that:

$$\hat{\varphi}_m(0)\hat{\varphi}_i(0) = \hat{\psi}_n(0)\hat{\psi}_j(0), \qquad [10.17a]$$

$$\hat{\varphi}_m(1)\hat{\varphi}_i(1) = \hat{\psi}_n(1)\hat{\psi}_j(1), \qquad [10.17b]$$

$$\frac{\partial\hat{\varphi}_m(0)}{\partial\xi}\frac{\partial\hat{\varphi}_i(0)}{\partial\xi} = \frac{\partial\hat{\psi}_n(0)}{\partial\eta}\frac{\partial\hat{\psi}_j(0)}{\partial\eta} \text{ and} \qquad [10.17c]$$

$$\frac{\partial\hat{\varphi}_m(1)}{\partial\xi}\frac{\partial\hat{\varphi}_i(1)}{\partial\xi} = \frac{\partial\hat{\psi}_n(1)}{\partial\eta}\frac{\partial\hat{\psi}_j(1)}{\partial\eta} \qquad [10.17d]$$

$m \backslash i$	$\hat{\varphi}_{i=1}$	$\hat{\varphi}_{i=2}$	$\hat{\varphi}_{i=3}$	$\hat{\varphi}_{i\geq 4}$
$\hat{\varphi}_{m=1}$	1	ξ	ξ^2	$\cos((i-3)\pi\xi)$
$\hat{\varphi}_{m=2}$		ξ^2	ξ^3	$\xi\cos((i-3)\pi\xi)$
$\hat{\varphi}_{m=3}$	SYMM		ξ^4	$\xi^2\cos((i-3)\pi\xi)$
$\hat{\varphi}_{m\geq 4}$				$\cos((m-3)\pi\xi)\cos((i-3)\pi\xi)$

Table 10.5. *Matrix showing the terms obtained multiplying the sets of admissible functions* $\hat{\varphi}_m(\xi)\hat{\varphi}_i(\xi)$

Thus, the matrices on the left hand side of equations [10.17a]–[10.17d] will be defined below which are required to compute the penalty matrix $\hat{\mathbf{P}}$. Then, the first step to define the terms in equations [10.17a]–[10.17b] used to model translational springs (or masses) is to write a matrix containing the products of the set of admissible functions shown in Table 10.5. Then evaluating the terms in Table 10.5 at $\xi = 0$ gives terms $\hat{\varphi}_m(0)\hat{\varphi}_i(0) = \hat{\psi}_n(0)\hat{\psi}_j(0)$ shown in Table 10.6.

$m \setminus i$	$\hat{\varphi}_{i=1}$	$\hat{\varphi}_{i=2}$	$\hat{\varphi}_{i=3}$	$\hat{\varphi}_{i\geq 4}$
$\hat{\varphi}_{m=1}$	1	0	0	1
$\hat{\varphi}_{m=2}$		0	0	0
$\hat{\varphi}_{m=3}$	SYMM		0	0
$\hat{\varphi}_{m\geq 4}$				1

Table 10.6. *Terms* $\hat{\varphi}_m(0)\hat{\varphi}_i(0) = \hat{\psi}_n(0)\hat{\psi}_j(0)$

Similarly, evaluating the terms in Table 10.5 at $\xi = 1$ gives terms $\hat{\varphi}_m(1)\hat{\varphi}_i(1) = \hat{\psi}_n(1)\hat{\psi}_j(1)$ shown in Table 10.7. Terms in the fourth row and column of Table 10.7 are either –1 or +1.

$m \setminus i$	$\hat{\varphi}_{i=1}$	$\hat{\varphi}_{i=2}$	$\hat{\varphi}_{i=3}$	$\hat{\varphi}_{i\geq 4}$
$\hat{\varphi}_{m=1}$	1	1	1	$\cos((i-3)\pi)$
$\hat{\varphi}_{m=2}$		1	1	$\cos((i-3)\pi)$
$\hat{\varphi}_{m=3}$	SYMM		1	$\cos((i-3)\pi)$
$\hat{\varphi}_{m\geq 4}$				$\cos((m-3)\pi)\ \cos((i-3)\pi)$

Table 10.7. *Terms* $\hat{\varphi}_m(1)\hat{\varphi}_i(1) = \hat{\psi}_n(1)\hat{\psi}_j(1)$

Then, matrices $\frac{\partial\hat{\varphi}_m(0)}{\partial\xi}\frac{\partial\hat{\varphi}_i(0)}{\partial\xi}$ and $\frac{\partial\hat{\varphi}_m(1)}{\partial\xi}\frac{\partial\hat{\varphi}_i(1)}{\partial\xi}$ to model rotational springs (or moments of inertia) can be built from a matrix showing the products of the first derivative of the set of admissible functions used in this work as shown in Table 10.3. Thus, evaluating the terms in Table 10.3 at $\xi = 0$ gives terms $\frac{\partial\hat{\varphi}_m(0)}{\partial\xi}\frac{\partial\hat{\varphi}_i(0)}{\partial\xi}$ shown in Table 10.8.

$m \backslash i$	$\hat{\varphi}'_{i=1}$	$\hat{\varphi}'_{i=2}$	$\hat{\varphi}'_{i=3}$	$\hat{\varphi}'_{i\geq 4}$
$\hat{\varphi}'_{m=1}$	0	0	0	0
$\hat{\varphi}'_{m=2}$		1	$2x$	0
$\hat{\varphi}'_{m=3}$		SYMM	$4x^2$	0
$\hat{\varphi}'_{m\geq 4}$				0

Table 10.8. *Terms* $\frac{\partial\hat{\varphi}_m(0)}{\partial\xi}\frac{\partial\hat{\varphi}_i(0)}{\partial\xi}$

Similarly, evaluating the terms in Table 10.3 at $\xi = 1$ gives terms $\frac{\partial\hat{\varphi}_m(1)}{\partial\xi}\frac{\partial\hat{\varphi}_i(1)}{\partial\xi}$ shown in Table 10.9.

$m \backslash i$	$\hat{\varphi}'_{i=1}$	$\hat{\varphi}'_{i=2}$	$\hat{\varphi}'_{i=3}$	$\hat{\varphi}'_{i\geq 4}$
$\hat{\varphi}'_{m=1}$	0	0	0	0
$\hat{\varphi}'_{m=2}$	0	1	2	0
$\hat{\varphi}'_{m=3}$	0	2	4	0
$\hat{\varphi}'_{m\geq 4}$	0	0	0	0

Table 10.9. *Terms* $\frac{\partial\hat{\varphi}_m(1)}{\partial\xi}\frac{\partial\hat{\varphi}_i(1)}{\partial\xi}$

10.4. Modes of vibration

The solution to the eigenproblem defined in equation [10.9] gives the eigenvalues of the system and their corresponding eigenvectors. The natural frequencies are the square root of the eigenvalues, while the eigenvectors define the vibration mode shape. When using the RRM the vibration modes of the continuum (whole structural element) are calculated using the eigenvectors. Thus, to calculate a mode of vibration of a plate as a function of the Cartesian coordinates it is necessary to carry out the summation of the set of admissible functions evaluated at the point of interest times their corresponding term of the eigenvector as shown in the following equation:

$$Mode_r(\xi,\eta) = X_{rs} \bullet \varphi_i(\xi)\psi_j(\eta), \qquad [10.18]$$

where the right hand side of the equation [10.18] is a dot product and the subscript r is the mode number that is calculated, while the subscript s defines the value of the row of the eigenvector which has to be multiplied by the functions $\varphi_i(\xi)\psi_j(\eta)$. Note that $s = j + (i-1)i_{\max}$.

10.5. Results

Table 10.10 shows the first six non-zero frequency parameters of the 21 cases of plates with classical boundary conditions (free, simply supported and clamped) obtained using the proposed set of admissible functions in the RRM and adding artificial springs to model constraints. The non-dimensional stiffness coefficients for the artificial springs (penalty parameters) were assigned a value equal to 10^{i-1}, for $i = 1,2...n$, while repeating the calculations n times till monotonic convergence was interrupted. The rigid body modes are not included in Table 10.10. Case 20 SFFF has one rigid body mode and case 21 FFFF has three rigid body modes. Results presented in this section were obtained using 40 terms in each direction.

Notice that the increment of the penalty parameters in this work is completely arbitrary and even better results could have been obtained if the values of the penalty parameters had been defined using smaller increments. This is due, as stated before, because there is a limit in the maximum value of the penalty parameters that will give the solution of the problem without numerical instabilities.

Case		Mode					
	$\hat{k}$	1	2	3	4	5	6
1 SSSS	1E+10	19.739	49.348	49.348	78.957	98.698	98.698
2 SCSC	1.E+09	28.951	54.744	69.329	94.589	102.220	129.105
3 SCSS	1.E+09	23.646	51.675	58.647	86.136	100.272	113.233
4 SCSF	1.E+09	12.687	33.065	41.702	63.016	72.399	90.614
5 SSSF	1.E+10	11.685	27.756	41.197	59.066	61.861	90.297
6 SFSF	1.E+09	9.631	16.135	36.726	38.945	46.739	70.741
7 CCCC	1.E+09	35.986	73.397	73.397	108.225	131.592	132.215
8 CCCS	1.E+09	31.827	63.333	71.079	100.798	116.363	130.361
9 CCCF	1.E+09	23.921	39.999	63.224	76.713	80.576	116.665
10 CCSS	1.E+09	27.054	60.540	60.787	92.840	114.562	114.709
11 CCSF	1.E+09	17.537	36.024	51.813	71.078	74.328	105.791
12 CCFF	1.E+09	6.920	23.905	26.585	47.653	62.708	65.535
13 CSCF	1.E+09	23.371	35.572	62.878	66.764	77.378	108.874
14 CSSF	1.E+09	16.792	31.114	51.397	64.022	67.541	101.117
15 CSFF	1.E+09	5.351	19.075	24.671	43.088	52.708	63.760
16 CFCF	1.E+09	22.168	26.407	43.596	61.177	67.179	79.818
17 CFSF	1.E+09	15.192	20.584	39.736	49.449	56.280	77.325
18 CFFF	1.E+09	3.471	8.507	21.285	27.199	30.957	54.188
19 SSFF	1.E+10	3.367	17.316	19.293	38.211	51.036	53.487
20 SFFF	1.E+09	6.644	14.902	25.376	26.001	48.450	50.579
21 FFFF		13.468	19.596	24.270	34.801	34.801	61.093

Table 10.10. *Frequency parameters of a plate with the 21 possible combinations of boundary conditions including free, simply supported and clamped conditions*

Case		Mode					
	$\hat{k}$	1	2	3	4	5	6
22 SSSG	1.E+09	12.337	32.076	41.946	61.685	71.555	91.296
23 SCSG	1.E+09	13.686	38.694	42.587	66.300	83.490	91.706
24 SGSF	1.E+09	9.736	17.685	39.189	42.384	47.967	74.526
25 SGSG	1.E+09	9.870	19.739	39.479	49.348	49.348	78.957
26 CSCG	1.E+09	23.816	39.090	63.537	75.843	79.528	114.785
27 CSSG	1.E+09	17.332	35.051	52.099	69.914	73.440	106.483
28 SSGG	1.E+09	4.935	24.674	24.674	44.413	64.153	64.153
29 CSGG	1.E+08	7.238	25.554	32.274	49.953	64.654	76.830
30 SSGF	1.E+09	4.034	18.821	24.010	41.174	53.026	63.287
31 SCGF	1.E+08	5.704	24.694	24.944	45.755	63.681	64.403
32 SGGF	1.E+09	2.408	9.181	21.997	30.510	33.426	56.190
33 SFGF	1.E+09	2.378	6.881	21.821	26.372	29.208	51.646
34 CGSG	1.E+09	15.418	23.646	49.966	51.675	58.647	86.136
35 CGCG	1.E+09	22.374	28.951	54.744	61.675	69.330	94.589
36 SGGG	1.E+08	2.467	12.337	22.207	32.076	41.946	61.685
37 CGGG	1.E+08	5.593	13.686	30.226	38.694	42.587	66.300
38 SGFG	1.E+08	11.685	15.418	27.756	41.197	49.965	59.066
39 CGFG	1.E+08	3.516	12.687	22.035	33.065	41.702	61.698
40 GGFG	1.E+07	5.593	9.736	17.685	30.226	39.188	42.384
41 FGFG	1.E+07	9.631	16.135	22.373	36.726	38.945	46.738
42 GGGG	1.E+07	9.870	9.870	19.739	39.478	39.478	49.348
43 CCCG	1.E+09	24.578	44.771	63.985	83.277	87.256	123.256

Table 10.11. *Frequency parameters of a plate with at least on edge under guided boundary conditions*

Table 10.11 shows the frequency parameters of the 34 cases of plates with classical boundary conditions (free, simply supported and clamped) and at least one edge of the plate under guided boundary conditions. These cases are identified as 22 to 55. Rigid body modes are again not included in Table 10.11. Cases 38 SGFG, 40 GGFG, 42 GGGG, 52 SGFF and 54 GGFF have one rigid body mode and cases 41 FGFG and 55 GFFF have two rigid body modes.

Case		Mode					
	$\hat{k}$	1	2	3	4	5	6
43 CCCG	1.E+09	24.578	44.771	63.985	83.277	87.256	123.256
44 CCSG	1.E+09	18.349	41.251	52.632	74.086	85.147	106.843
45 CCGF	1.E+09	7.776	25.850	32.217	51.192	64.917	76.337
46 CCGG	1.E+07	8.996	32.895	33.051	55.008	77.226	77.291
47 CGCF	1.E+09	22.259	27.495	48.533	61.402	68.199	90.289
48 CGSF	1.E+09	15.293	21.897	45.058	49.684	57.400	82.031
49 CSGF	1.E+09	6.601	19.954	31.677	47.034	53.632	76.003
50 CGGF	1.E+08	5.541	10.898	30.024	34.223	37.326	61.183
51 CFGF	1.E+08	5.508	8.986	27.359	29.857	36.177	56.973
52 SGFF	1.E+08	8.700	15.273	26.365	32.867	49.568	53.854
53 CGFF	1.E+08	3.493	10.181	21.838	31.427	34.029	58.071
54 GGFF	1.E+07	4.899	6.068	15.922	29.277	30.611	40.376
55 GFFF	1.E+07	5.366	14.621	22.002	29.681	36.045	40.050

Table 10.11. (continued) *Frequency parameters of a plate with at least on edge under guided boundary conditions*

A comparison of the results in Tables 10.10 and 10.11 against the results by Bert and Malik [BER 94] revealed that the maximum difference between the two sets of results was always found in the third decimal place.

Cases with two opposite edges, either simply supported or guided, have an analytical solution, while the solution of the remaining cases can be solved using approximate or numerical methods [BER 94]. Furthermore, Bert and Malik classified the 55 cases presented in Tables 10.10 and 10.11 according to the boundary conditions and type of solutions:

– cases 1 to 6 with two opposite edges simply supported have an analytical solution;

– cases 7 to 21 can be solved by approximate or numerical methods only. However in a more recent publications exact solution for cases 7 CCCC, 8 CCCS and 10 CCSS have been found in [WU 07] and [YUF 09];

– cases 22 to 25 with two opposite edges simply supported have an analytical solution;

– cases 26 to 33 with one edge simply supported and opposite edge guided have an analytical solution;

– cases 34 to 42 with two opposite edges guided have an analytical solution;

– cases 43 to 55 can be solved by approximate or numerical methods only.

The first three modes are rigid body modes (frequency parameter is equalt to zero). Rigid body modes are not included in Tables 10.10 and 10.11. The first three modes are rigid body modes (one translation and two rotational modes corresponding to zero frequencies) and are presented in Figure 10.2. The following six modes, corresponding to the frequency parameters, given in Table 10.6.1 are shown in Figure 10.3.

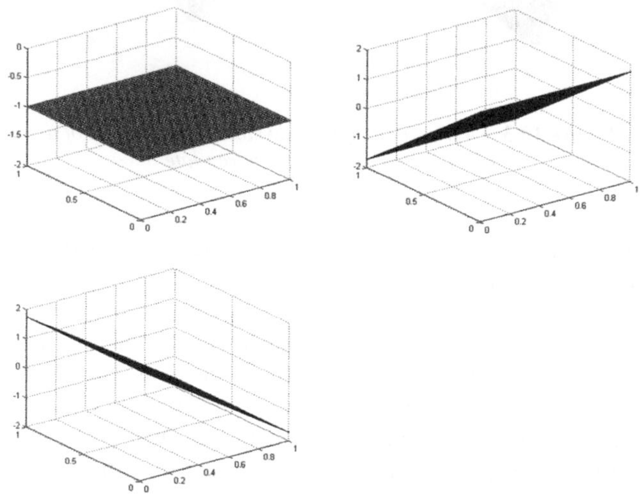

Figure 10.2. *First three modes of vibration (rigid body modes) of a FFFF (unconstrained) square plate*

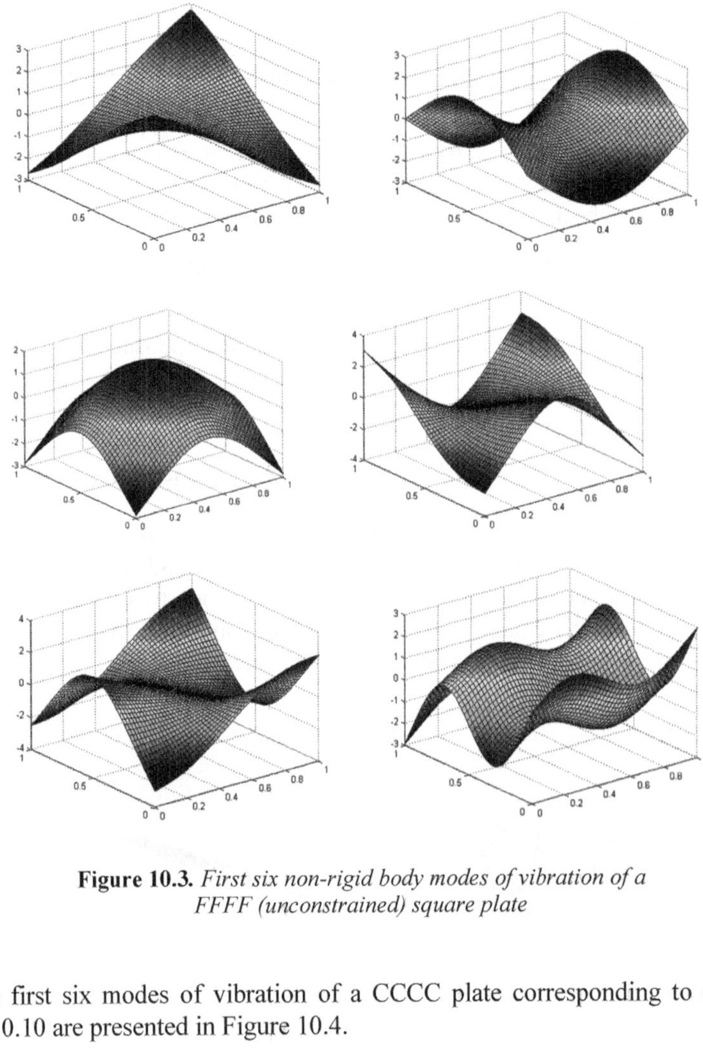

Figure 10.3. *First six non-rigid body modes of vibration of a FFFF (unconstrained) square plate*

The first six modes of vibration of a CCCC plate corresponding to case 7 in Table 10.10 are presented in Figure 10.4.

That then ends our story about natural frequencies and modes of thin rectangular plates using the RRM with a nice set of admissible functions. Nice in the sense they are easy to work with and behave reasonably well numerically. While the procedure allows calculating as many frequencies of a completely free plate as the computer memory would permit, the computations for higher modes will be of no use as the thin plate assumption would not hold for higher modes. We must also note here that for the sort of plate problems we have discussed, analytical exact solutions or Gorman's superposition method may give more accurate results but the advantage of the RRM is that any variation to the system, such as the addition of a mass, or an intermediate support, could be easily achieved.

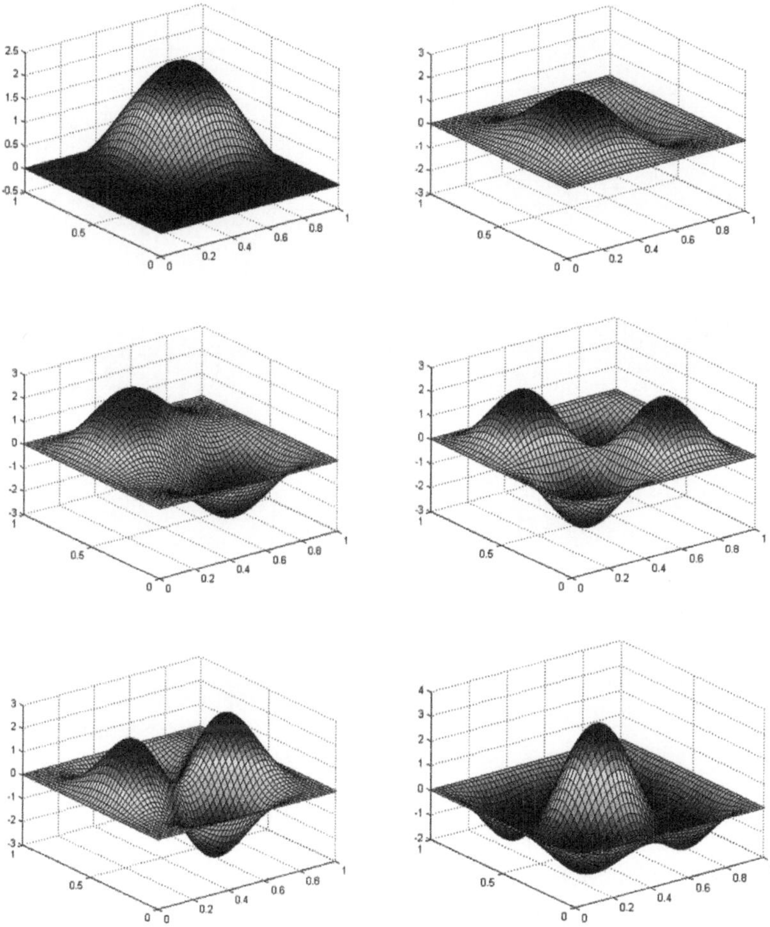

Figure 10.4. *First six modes of vibration of a CCCC square plate*

11

Natural Frequencies and Modes of Shallow Shells of Rectangular Planform

The method presented in this chapter (as in the previous chapters for beams and plates) uses the RRM with a set of admissible functions to model the deflected shape of an unconstrained structure. Thus, in general, the first step of the solutions of the vibration problem is the computation of the stiffness and mass matrices of an unconstrained shallow shell obtained from the Rayleigh–Ritz minimization. Then the penalty method through the use of artificial springs (or masses) is used to add constraints to the shells. The same approach can be found in [MON 09] and a similar approach can be found in the work by Crossland and Dickinson [CRO 97] to model shells with slits. In [CRO 97] Crossland and Dickinson used simple polynomials to define the sets of admissible functions of each free shell element and artificial springs were used to join the shell elements as well as to define boundary conditions.

11.1. Theoretical derivations of the eigenvalue problems

In this chapter, a Rayleigh–Ritz procedure to obtain frequency parameters of thin shallow shells of rectangular planform is presented. The shells are assumed to have isotropic material with Young's modulus E, Poisson ratio ν and density ρ. The geometric properties of the shallow shells of constant thickness h are defined placing the midpoint of the middle surface of the shell at $z=0$, one corner of the middle surface of the shell at $x=0,\ y=0$ and aligning the edges of the shells with the x and y axes. The projections of the edges on their respective axis are defined as length a and width b. The middle surface of the shell is defined by two constant curvatures of radii R_x and R_y, which are parallel to the x and y axes. The radii R_x and R_y have their origins located on points along a line parallel to axis z which crosses the midpoint of the shell located at point $x = a/2$, $y = b/2$ as shown in Figure 11.1. Thus, the middle surface of the shell is described by:

$$z = -\frac{\left(x - \frac{a}{2}\right)^2}{2R_x} - \frac{\left(y - \frac{b}{2}\right)^2}{2R_y} \qquad [11.1]$$

The geometry of the shallow shells can be classified according to the ratios of the length of the sides to their radii of curvature as:

– plate ($a/R_x = 0$ and $b/R_y = 0$),

– cylindrical ($a/R_x = 0$ and $b/R_y \neq 0$),

– spherical ($a/R_x = b/R_y \neq 0$) and

– hyperbolic paraboloidal ($a/R_x \neq 0$ and $-b/R_y \neq 0$).

Frequency parameters of shallow shells with other geometries can also be obtained using the procedure described here.

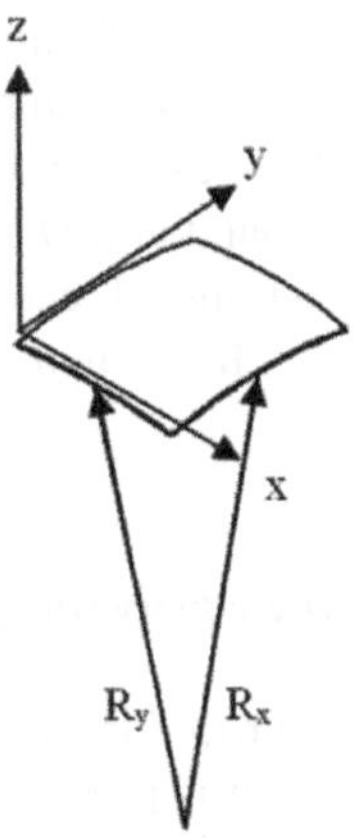

Figure 11.1. *Co-ordinate axes of shallow-shells*

Presenting the procedure in a way as in previous chapters, the first step is to present the sets of admissible functions needed to describe the deflection shape of the shallow shells. In this case, the sets of admissible functions are exactly the same as for plates. These sets of admissible functions are presented again below for completion.

Set of admissible functions in the x ***direction***

$$\phi_i(x) = 1 \text{ for } i = 1 \qquad [11.2a]$$

$$\phi_i(x) = \frac{x}{L} \text{ for } i = 2 \qquad [11.2b]$$

$$\phi_i(x) = \left(\frac{x}{L}\right)^2 \text{ for } i = 3 \qquad [11.2c]$$

$$\phi_i(x) = \cos\frac{(i-3)\pi x}{a} \text{ for } i \geq 4 \qquad [11.2d]$$

Set of admissible functions in the y ***direction***

$$\psi_j(y) = 1 \text{ for } j = 1 \qquad [11.2e]$$

$$\psi_j(y) = \frac{y}{L} \text{ for } j = 2 \qquad [11.2f]$$

$$\psi_j(y) = \left(\frac{y}{L}\right)^2 \text{ for } j = 3 \qquad [11.2g]$$

$$\psi_j(y) = \cos\frac{(j-3)\pi y}{b} \text{ for } j \geq 4 \qquad [11.2h]$$

or in non-dimensional form.

Non-dimensional set of admissible functions in the x ***direction***

$$\hat{\phi}_i(\xi) = 1 \text{ for } i = 1 \qquad [11.3a]$$

$$\hat{\phi}_i(\xi) = \xi \text{ for } i = 2 \qquad [11.3b]$$

$$\hat{\phi}_i(\xi) = \xi^2 \text{ for } i = 3 \qquad [11.3c]$$

$$\hat{\phi}_i(\xi) = \cos(i-3)\pi\xi \text{ for } i \geq 4 \quad [11.3d]$$

where: $\xi = x / a$, and

Non-dimensional set of admissible functions in the y direction

$$\hat{\psi}_j(y) = 1 \text{ for } j = 1 \quad [11.3e]$$

$$\hat{\psi}_j(y) = \eta \text{ for } j = 2 \quad [11.3f]$$

$$\hat{\psi}_j(y) = \eta^2 \text{ for } j = 3 \quad [11.3g]$$

$$\psi_j(y) = \cos(j-3)\pi\eta \text{ for } j \geq 4 \quad [11.3h]$$

with $\eta = y / b$.

For convenience, the same number of functions N will be used in both directions x and y in the displacement amplitude functions $U(x,y)$, $\mathrm{V}(x,y)$ and $W(x,y)$ defined as

$$U(x,y) = \sum_{i=1}^{N}\sum_{j=1}^{N} a_{ij}\phi_i(x)\psi_j(y), \quad [11.4a]$$

$$V(x,y) = \sum_{i=1+N}^{2N}\sum_{j=1+N}^{2N} a_{ij}\phi_i(x)\psi_j(y), \quad [11.4b]$$

$$U(x,y) = \sum_{i=1}^{3N}\sum_{j=1}^{3N} a_{ij}\phi_i(x)\psi_j(y), \quad [11.4c]$$

or in non-dimensional form

$$\hat{U}(\xi,\eta) = \sum_{i=1}^{N}\sum_{j=1}^{N} a_{ij}\hat{\phi}_i(\xi)\hat{\psi}_j(\eta), \quad [11.4a]$$

$$\hat{V}(x,y) = \sum_{i=1+N}^{2N}\sum_{j=1+N}^{2N} a_{ij}\hat{\phi}_i(\xi)\hat{\psi}_j(\eta), \quad [11.4b]$$

$$\hat{W}(x,y)=\sum_{i=1+2N}^{3N}\sum_{j=1+2N}^{3N} a_{ij}\hat{\phi}_i(\xi)\hat{\psi}_j(\eta), \qquad [11.4c]$$

where a_{ik} are unknown coefficients.

Therefore, the displacement of the middle surface of the shallow shell in the x, y and z directions is given by:

$$u(x,y)=U(x,y)\sin\omega t, \qquad [11.5a]$$

$$v(x,y)=V(x,y)\sin\omega t \text{ and} \qquad [11.5b]$$

$$w(x,y)=W(x,y)\sin\omega t \qquad [11.5c]$$

The Rayleigh–Ritz procedure to compute the stiffness matrix and the mass matrix of the shallow shells consists of substituting the displacement amplitude functions $U(x,y)$, $\mathrm{V}(x,y)$ and $W(x,y)$ in the energy terms and carrying out the Rayleigh–Ritz minimization of the energy terms with respect to the unknown coefficients a_{ij}.

The maximum potential energy of a shallow shell due to stretching and bending as defined by Leissa and Narita [LEI 84] is:

$$\begin{aligned}V_{\max} &= \frac{Eh}{2(1-\nu^2)}\int_0^b\int_0^a\left[(\varepsilon_x+\varepsilon_y)^2-2(1-\nu)(\varepsilon_x\varepsilon_y-\gamma_{xy}/4)\right]dxdy\\ &+\frac{Eh^3}{24(1-\nu^2)}\int_0^b\int_0^a\left[(\kappa_x+\kappa_y)^2-2(1-\nu)(\kappa_x\kappa_y-\kappa_{xy}^2)\right]dxdy\end{aligned} \qquad [11.6]$$

where ε_x and ε_y are the membrane lengthening strains and γ_{xy} is the shift strain, while κ_x, κ_y and κ_z are the curvature changes of the shell. These strains and curvatures are defined as [LEI 84]:

$$\varepsilon_x=\partial u/\partial x+w/R_x, \qquad [11.7a]$$

$$\varepsilon_y=\partial v/\partial y+w/R_y, \qquad [11.7b]$$

$$\gamma_{xy} = \partial v / \partial x + \partial u / \partial y \,, \tag{11.7c}$$

$$\kappa_x = \partial^2 w / \partial x^2 \,, \tag{11.7d}$$

$$\kappa_y = \partial^2 w / \partial y^2 \text{ and} \tag{11.7e}$$

$$\kappa_{xy} = \partial^2 w / \partial x \partial y \tag{11.7f}$$

The maximum kinetic energy of the shell is [LEI 84]

$$T_{\max} = \frac{\rho h}{2} \int_0^b \int_0^a \left[\left(\frac{\partial u}{\partial t} \right)^2 + \left(\frac{\partial v}{\partial t} \right)^2 + \left(\frac{\partial w}{\partial t} \right)^2 \right] dx dy \tag{11.8}$$

Applying the Rayleigh–Ritz minimization using non-dimensional coordinates gives the stiffness matrix and the mass matrix to be used in the following generalized eigenproblem

$$\left(\frac{D}{ab} \hat{\mathbf{K}} - \lambda^2 abh\rho \hat{\mathbf{M}} \right) \{\mathbf{a}\} = 0 \,, \tag{11.9a}$$

or in non-dimensional form

$$\left(\hat{\mathbf{K}} - \lambda^2 \hat{\mathbf{M}} \right) \{\mathbf{a}\} = 0 \,, \tag{11.9b}$$

where $\hat{\mathbf{K}}$ is the non-dimensional stiffness matrix, $\hat{\mathbf{M}}$ is the non-dimensional mass matrix, λ is the eigenvalue and $\{\mathbf{a}\}$ is an eigenvector. In this case the circular natural frequencies in radians per second are the square roots of the eigenvalues and the eigenvectors can be used to compute the vibration modes when substituted in the displacement functions given in [11.4]. The non-dimensional stiffness and mass matrices are defined as follows:

$$\hat{\mathbf{K}} = \begin{bmatrix} \hat{\mathbf{K}}_{uu} & \hat{\mathbf{K}}_{uv} & \hat{\mathbf{K}}_{uw} \\ & \hat{\mathbf{K}}_{vv} & \hat{\mathbf{K}}_{vw} \\ sym. & & \hat{\mathbf{K}}_{ww} \end{bmatrix} \tag{11.10a}$$

where

$$\hat{\mathbf{K}}_{uu} = \left(\frac{b}{h}\right)^2 E_{mi}^{(1,1)} F_{nj}^{(0,0)} + \frac{(1-\nu)}{2}\left(\frac{a}{h}\right)^2 E_{mi}^{(0,0)} F_{nj}^{(1,1)} \qquad [11.10b]$$

$$\hat{\mathbf{K}}_{uv} = \nu\left(\frac{a}{h}\right)\left(\frac{b}{h}\right) E_{mi}^{(1,0)} F_{nj}^{(0,1)} + \frac{(1-\nu)}{2}\left(\frac{a}{h}\right)\left(\frac{b}{h}\right) E_{mi}^{(0,1)} F_{nj}^{(1,0)} \qquad [11.10c]$$

$$\hat{\mathbf{K}}_{uw} = \left(\frac{b}{R_x} + \frac{\nu b}{R_y}\right)\left(\frac{a}{h}\right)\left(\frac{b}{h}\right) E_{mi}^{(1,0)} F_{nj}^{(0,0)} \qquad [11.10d]$$

$$\hat{\mathbf{K}}_{vv} = \left(\frac{a}{h}\right)^2 E_{mi}^{(0,0)} F_{nj}^{(1,1)} + \frac{(1-\nu)}{2}\left(\frac{b}{h}\right)^2 E_{mi}^{(1,1)} F_{nj}^{(0,0)} \qquad [11.10e]$$

$$\hat{\mathbf{K}}_{vw} = \left(\frac{\nu a}{R_x} + \frac{a}{R_y}\right)\left(\frac{a}{h}\right)\left(\frac{b}{h}\right) E_{mi}^{(0,0)} F_{nj}^{(1,0)} \qquad [11.10f]$$

$$\begin{aligned}\hat{\mathbf{K}}_{ww} = {} & \left[\left(\frac{b}{R_x}\right)^2 + 2\nu\left(\frac{b}{R_x}\right)\left(\frac{b}{R_y}\right) + \left(\frac{b}{R_y}\right)^2\right]\left(\frac{a}{h}\right)^2 E_{mi}^{(0,0)} F_{nj}^{(0,0)} \\ & + \frac{1}{12}\left(\frac{b}{a}\right)^2 E_{mi}^{(2,2)} F_{nj}^{(0,0)} + \frac{1}{12}\left(\frac{a}{b}\right)^2 E_{mi}^{(0,0)} F_{nj}^{(2,2)} \\ & + \frac{\nu}{12} E_{mi}^{(0,2)} F_{nj}^{(2,0)} + \frac{\nu}{12} E_{mi}^{(2,0)} F_{nj}^{(0,2)} + \frac{2(1-\nu)}{12} E_{mi}^{(1,1)} F_{nj}^{(1,1)}\end{aligned} \qquad [11.10g]$$

$$\hat{\mathbf{M}} = \begin{bmatrix} \hat{\mathbf{M}}_{uu} & 0 & 0 \\ & \hat{\mathbf{M}}_{vv} & 0 \\ sym. & & \hat{\mathbf{M}}_{ww} \end{bmatrix} \qquad [11.11a]$$

$$\hat{\mathbf{M}}_{uu} = \hat{\mathbf{M}}_{vv} = \hat{\mathbf{M}}_{ww} = E_{mi}^{(0,0)} F_{nj}^{(0,0)} \qquad [11.11b]$$

where

$$E_{mi}^{(r,s)} = \int_0^1 \left(\frac{d^r \hat{\varphi}_n}{d\xi^m} \right) \left(\frac{d^s \hat{\varphi}_j}{d\xi^i} \right) d\xi \text{ and } F_{nj}^{(r,s)} = \int_0^1 \left(\frac{d^r \hat{\psi}_n}{d\eta^r} \right) \left(\frac{d^s \hat{\psi}_j}{d\eta^s} \right) d\eta$$

Solving the non-dimensional generalized eigenproblem in [11.9b] gives results for the non-dimensional frequency parameters λ defined as:

$$\lambda = \omega ab\sqrt{ph / D} \qquad [11.12]$$

The terms $E_{mi}^{(0,0)}$, $E_{mi}^{(1,1)}$, $E_{mi}^{(0,2)}$ and $E_{mi}^{(2,2)}$ were already computed in previous chapters for beams and plates. Furthermore, remembering that for non-dimensional matrices $E_{mi}^{(r,s)} = F_{nj}^{(r,s)}$ and $E_{mi}^{(r,s)} = E_{im}^{(s,r)\mathrm{T}}$ where T denotes the transpose of a matrix, the only matrix that needs to be computed is $E_{mi}^{(1,0)}$. Following the same procedure as in previous chapters, the first step to compute all terms $E_{mi}^{(1,0)}$ is to gather a matrix that shows the products of the first derivative $\hat{\phi}'$ (rows) and the functions $\hat{\phi}$ (columns) as shown in Table 11.1.

$m \backslash i$	$\hat{\phi}_{i=1}$	$\hat{\phi}_{i=2}$	$\hat{\phi}_{i=3}$	$\hat{\phi}_{i\geq 4}$
$\hat{\phi}'_{m=1}$	0	0	0	0
$\hat{\phi}'_{m=2}$	1	ξ	ξ^2	$\cos((i-3)\pi\xi)$
$\hat{\phi}'_{m=3}$	2ξ	$2\xi^2$	$2\xi^3$	$2\xi\cos((i-3)\pi\xi)$
$\hat{\phi}'_{m\geq 4}$	$-(m-3)\pi$ $\sin((m-3)\pi\xi)$	$-(m-3)\pi\xi$ $\sin((m-3)\pi\xi)$	$-(m-3)\pi\xi^2$ $\sin((m-3)\pi\xi)$	$-(m-3)\pi$ $\sin((m-3)\pi\xi)$ $\cos((i-3)\pi\xi)$

Table 11.1. *Products of the first derivatives of the set of admissible functions $\hat{\phi}'_m$ and the set of admissible functions $\hat{\phi}_i$ to build matrix $E_{mi}^{(1,0)}$*

The second step is to carry out the integrations of the terms in Table 11.1 with respect to ξ with limits of integration 0 and 1 as shown in Table 11.2.

$m \backslash i$	$\hat{\phi}_{i=1}$	$\hat{\phi}_{i=2}$	$\hat{\phi}_{i=3}$	$\hat{\phi}_{i\geq 4}$
$\hat{\phi}'_{m=1}$	0	0	0	0
$\hat{\phi}'_{m=2}$	1	1/2	1/3	0
$\hat{\phi}'_{m=3}$	1	2/3	1/2	$\int \hat{\phi}'_{m=3}\hat{\phi}_{i\geq 4} d\xi$
$\hat{\phi}'_{m\geq 4}$	$\int \hat{\phi}'_{m\geq 4}\hat{\phi}_{i=1} d\xi$	$\int \hat{\phi}'_{m\geq 4}\hat{\phi}_{i=2} d\xi$	$\int \hat{\phi}'_{m\geq 4}\hat{\phi}_{i=3} d\xi$	$\int \hat{\phi}'_{m\geq 4}\hat{\phi}_{i\geq 4} d\xi$

Table 11.2. *Terms* $E_{mi}^{(1,0)}$

The terms defined by integrals in Table 11.2 are:

$$\int \hat{\phi}'_{m=3}\hat{\phi}_{i\geq 4} d\xi = \frac{2\left(\cos\left((i-3)\pi\right)-1\right)}{(i-3)^2 \pi^2},$$

$$\int \hat{\phi}'_{m\geq 4}\hat{\phi}_{i=1} d\xi = \cos\left((m-3)\pi\right)-1,$$

$$\int \hat{\phi}'_{i=4}\hat{\phi}_{m\geq 2} d\xi = \cos\left((m-3)\pi\right),$$

$$\int \hat{\phi}'_{m\geq 4}\hat{\phi}_{i=3} d\xi = \cos\left((m-3)\pi\right)-\frac{2\left(\cos\left((m-3)\pi\right)-1\right)}{(m-3)^2 \pi^2},$$

and the last term of the matrix in Table 11.2 is:

$$\int \hat{\phi}'_{m\geq 4}\hat{\phi}_{i\geq 4} d\xi = \left((m-3)/2\right)\frac{\cos\left((m-i)\pi\right)-1}{m-i}+\frac{\cos\left((m+i)\pi\right)-1}{m+i-6},$$

Thus $\int \hat{\phi}'_{m\geq 4}\hat{\phi}_{i\geq 4} d\xi = 0$ if $m = i$ or if $m+i$ is equal to an even number. Furthermore, terms $E_{m\geq 4, i\geq 4}^{(1,0)}$ are positive if they are below the main diagonal and negative if they are above the main diagonal.

11.2. Frequency parameters of constrained shallow shells

Any combination of boundary conditions containing constraints is modeled through the use of artificial springs and/or artificial masses. Constraints can be defined at the edges of the shell as well as at locations in the interior of the shell.

A simply support along an edge is defined by constraining the in-plane displacements U and V, as well as the out of plane displacement W. A clamped support along an edge is completed adding a rotational spring to enforce zero normal slope W'. For instance, the potential energy terms needed to constrain a clamped edge at $\xi = 0$ using artificial spring in a similar way as described by Young and Dickinson [YOU 95] are

$$V_{transu} = \frac{1}{2}\hat{p}\int_0^1 \left(\left(\hat{U} \right)_{\xi=0} \right)^2 d\eta ,$$

$$V_{transv} = \frac{1}{2}\hat{p}\int_0^1 \left(\left(\hat{V} \right)_{\xi=0} \right)^2 d\eta ,$$

$$V_{transw} = \frac{1}{2}\hat{p}\int_0^1 \left(\left(\hat{W} \right)_{\xi=0} \right)^2 d\eta ,$$

$$V_{rotw} = \frac{1}{2}\hat{p}\int_0^1 \left(\left(\partial \hat{W} / \partial \xi \right)_{\xi=0} \right)^2 d\eta ,$$

where the non-dimensional penalty parameter $\hat{p}$ represents the non-dimensional stiffness of the translational and rotational springs defined as:

$$\hat{p} = \frac{ka^3}{D} = \frac{k_r a}{D} ,$$

where k and k_r are the stiffness coefficient of the springs (assuming equal values for springs at different locations).

If artificial masses are used to model constraints, the penalty matrix must be added to the mass matrix instead of to the stiffness matrix and the non-dimensional penalty parameter is defined as:

$$\alpha = \frac{m_u}{ma} = \frac{I_w}{ma^3} ,$$

where m_u and I_w are the values of a mass acting in an x direction and a moment of inertia around axis z (assuming equal values for mass properties in all directions), while m is the mass of the shell per unit area. Thus the penalty matrix has the following shape:

$$\hat{\mathbf{P}} = \begin{bmatrix} \hat{\mathbf{P}}_{uu} & 0 & 0 \\ & \hat{\mathbf{P}}_{vv} & 0 \\ sym. & & \hat{\mathbf{P}}_{ww} \end{bmatrix}, \qquad [11.13]$$

where the submatrices $\hat{\mathbf{P}}_{uu}$, $\hat{\mathbf{P}}_{vv}$ and $\hat{\mathbf{P}}_{ww}$ contain the constraints in the x, y and z directions. These submatrices are identical to the penalty matrices of a plate.

11.3. Results and discussion

The procedure described in this chapter enables us to obtain the natural frequencies and modes of vibration of any thin shallow-shell of rectangular planform using a single computer code. This is achieved choosing the appropriate geometric and materials properties, as well as activating the penalty parameters that define the desired boundary conditions. For instance, in the Matlab code, values for R_x and R_y can be defined with an infinite value to model plates.

Results presented in this chapter correspond to spherical shallow shells of rectangular planform with different boundary conditions, which will be identified by F, S and C for free, simply-supported and clamped edges, respectively. Four letters are necessary to fully identify the boundary conditions of the shells and they are given in the consecutive order with respect to the location of the edges where the constraints are modeled. Results presented in Table 11.3 for a FFFF shell presented here match the results presented by Qatu and Asadi [QAT 12], except for the fourth frequency parameter for which our result is 0.01 lower.

Table 11.4 presents results for spherical shells ($R_x / R_y = 1$) with different a / R_x ratios. These sets of results are used for comparison against the results presented by Mochida *et al.* for shallow shells clamped along the four edges. All cases presented in Table 11.4 have the following geometric and material characteristics $a / b = 1$, $a / h = 100$ and $\nu = 0.3$.

Case		Mode					
	$\hat{p}$	1	2	3	4	5	6
1 SSSS	10^{10}	27.99	51.52	51.52	80.40	99.64	99.79
2 SCSC	10^{9}	34.69	56.62	70.78	95.71	103.2	129.9
3 SCSS	10^{10}	30.63	53.72	60.42	87.41	101.3	114.1
4 SCSF	10^{10}	18.41	36.53	42.44	64.40	73.71	91.16
5 SSSF	10^{10}	18.07	31.94	42.04	60.58	63.41	90.88
6 SFSF	10^{10}	16.53	20.26	39.20	39.64	47.24	71.84
7 CCCC	10^{9}	40.42	74.71	74.71	109.1	132.2	133.1
8 CCCS	10^{9}	36.94	64.91	72.47	101.8	117.2	131.1
9 CCCF	10^{9}	26.91	42.74	63.64	77.91	81.56	117.4
10 CCSS	10^{9}	33.13	62.23	62.48	93.98	115.3	115.7
11 CCSF	10^{9}	21.74	39.14	52.37	72.25	75.59	106.3
12 CCFF	10^{8}	7.891	23.89	27.92	49.25	62.96	66.22
13 CSCF	10^{9}	26.62	38.72	63.35	68.14	78.44	109.7
14 CSSF	10^{9}	21.42	34.80	52.03	65.49	68.81	102.0
15 CSFF	10^{9}	6.87	19.57	25.94	44.91	53.29	64.34
16 CFCF	10^{10}	16.53	20.26	39.20	39.64	47.24	71.84
17 CFSF	10^{9}	19.93	23.73	41.99	49.97	56.63	78.45
18 CFFF	10^{9}	3.753	8.487	21.52	28.26	30.51	44.09
19 SSFF	10^{10}	5.872	17.46	21.50	40.35	51.48	55.45
20 SFFF	10^{11}	1.611	6.637	15.51	25.05	27.10	44.09
21 FFFF	N/A	13.46	19.56	25.99	34.84	34.84	61.77

Table 11.3. *Frequency parameters of a square spherical shells with the 21 possible combinations of boundary conditions including free, simply supported and clamped conditions*

			Mode					
	$\hat{p}$	a/R_x	1	2,3	4	5	6	7,8
Present	10^9	0.1	58.29	81.76	114.2	136.0	137.7	168.8
[MOC 12]			58.30	81.75	114.1	136.0	137.7	168.7
			Mode					
	$\hat{p}$	a/R_x	1,2	3	4	5	6,7	8
Present	10^9	0.3	130.2	134.0	153.4	167.2	181.1	196.4
[MOC 12]			130.2	134.0	153.4	167.2	181.2	196.3
			Mode					
	$\hat{p}$	a/R_x	1	2,3	4	5	6,7	8
Present	10^{10}	0.5	192.0	196.9	210.0	216.2	242.2	257.4
[MOC 12]			192.0	196.9	210.0	216.2	242.2	257.4

Table 11.4. *Frequency parameters of square spherical shells with clamped boundary conditions along its four edges*

The modes of vibration for the case where $a/R_x = 0.5$ in Table 11.4 are presented in Figures 11.2(a)–(h).

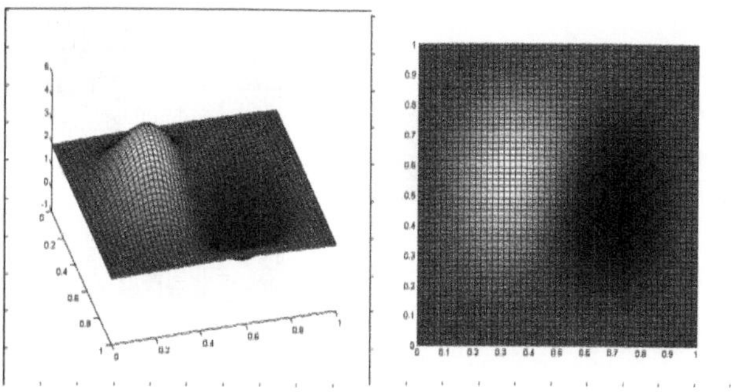

Figure 11.2(a). *First mode of vibration of shallow-shells with* $a/R_x = 0.5$

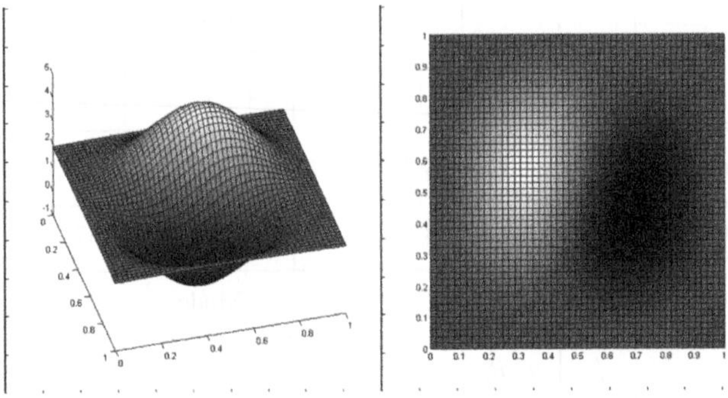

Figure 11.2(b). *Second mode of vibration of shallow-shells with* $a / R_x = 0.5$

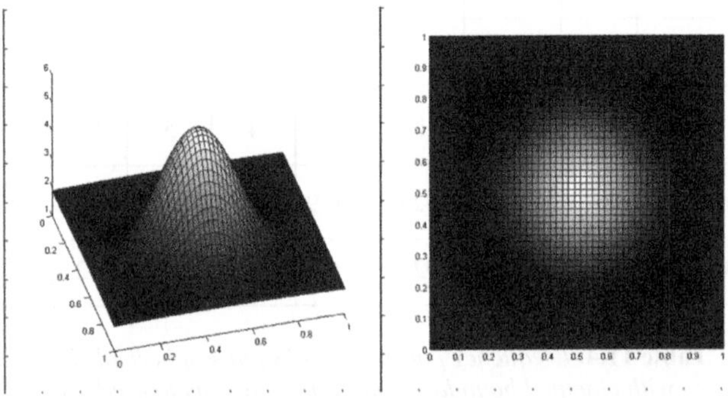

Figure 11.2(c). *Third mode of vibration of shallow-shells with* $a / R_x = 0.5$

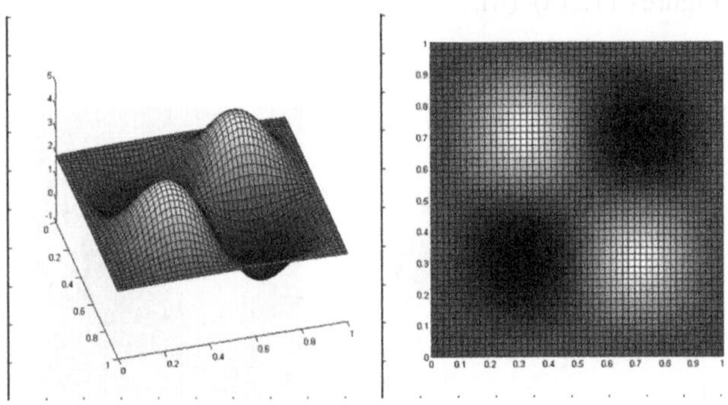

Figure 11.2(d). *Fourth mode of vibration of shallow-shells with* $a / R_x = 0.5$

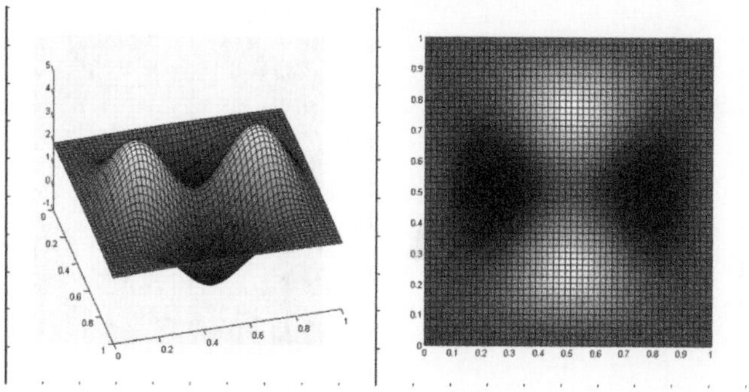

Figure 11.2(e). *Fifth mode of vibration of shallow-shells with* $a / R_x = 0.5$

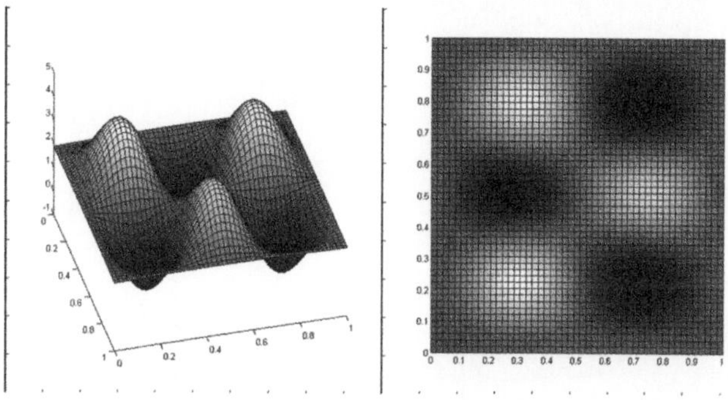

Figure 11.2(f). *Sixth mode of vibration of shallow-shells with* $a / R_x = 0.5$

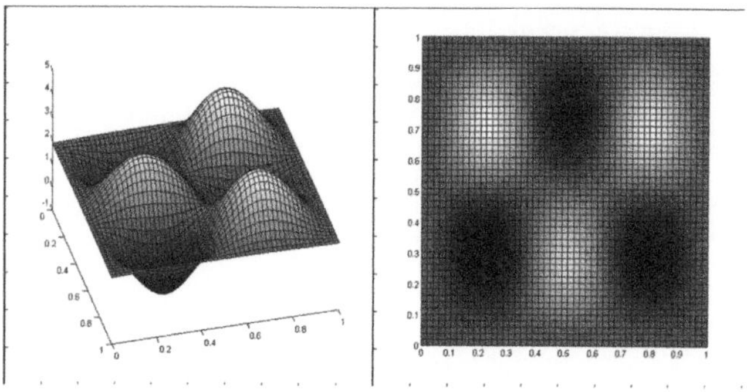

Figure 11.2(g). *Seventh mode of vibration of shallow-shells with* $a / R_x = 0.5$

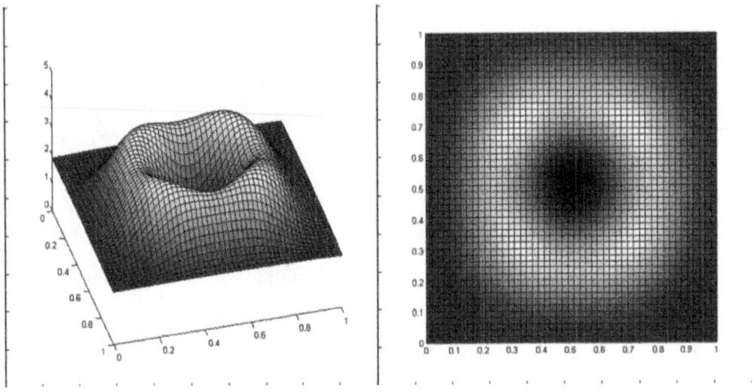

Figure 11.2(h). *Eight mode of vibration of shallow-shells with* $a / R_x = 0.5$

12

Natural Frequencies and Modes of Three-dimensional Bodies

The Rayleigh–Ritz procedure presented in the previous chapters is used to solve vibration problems of three-dimensional (3D) bodies. As in the previous chapters, the procedure uses admissible functions of an unconstrained body and artificial stiffness or artificial masses are used to add constraints to the 3D body.

12.1. Theoretical derivations of the eigenvalue problems

Consider a 3D body of rectangular planform with its edges aligned with the three axes of a Cartesian coordinate system, having one corner at the origin and the opposite corner at positive values a, b and h along the x, y and z axes as shown in Figure 12.1. The structure is originally unconstrained and as in the previous chapters artificial springs or artificial masses are added to model constraints.

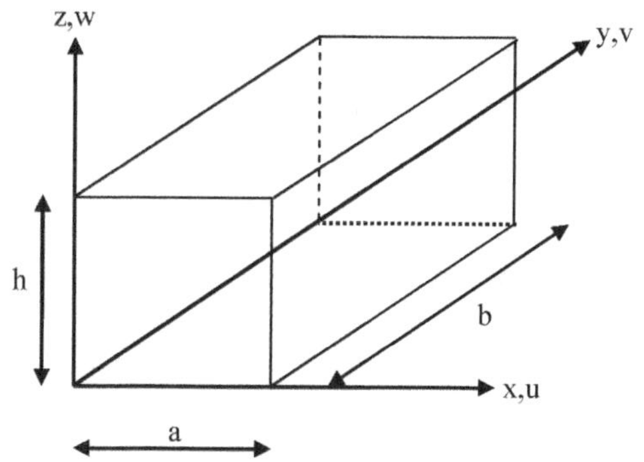

Figure 12.1. *Three-dimensional body displacements and Cartesian coordinate system*

Presenting the procedure in the same way as in previous chapters, the first step is to define sets of admissible functions that describe the deflection shape of the 3D body. In this case, sets of admissible functions in the x, y and z are needed.

Set of admissible functions in the x direction

$$\phi_i(x) = 1 \text{ for } i = 1 \tag{12.1a}$$

$$\phi_i(x) = \frac{x}{a} \text{ for } i = 2 \tag{12.1b}$$

$$\phi_i(x) = \left(\frac{x}{a}\right)^2 \text{ for } i = 3 \tag{12.1c}$$

$$\phi_i(x) = \cos\frac{(i-3)\pi x}{a} \text{ for } i \geq 4 \tag{12.1d}$$

Set of admissible functions in the y direction

$$\psi_j(y) = 1 \text{ for } j = 1 \tag{12.1e}$$

$$\psi_j(y) = \frac{y}{b} \text{ for } j = 2 \tag{12.1f}$$

$$\psi_j(y) = \left(\frac{y}{b}\right)^2 \text{ for j} = 3 \tag{12.1g}$$

$$\psi_j(y) = \cos\frac{(j-3)\pi y}{b} \text{ for j} \geq 4 \tag{12.1h}$$

Set of admissible functions in the z direction

$$\gamma_k(z) = 1 \text{ for } k = 1 \tag{12.1i}$$

$$\gamma_k(y) = \frac{z}{h} \text{ for } k = 2 \tag{12.1j}$$

$$\gamma_k(y) = \left(\frac{y}{h}\right)^2 \quad \text{for } k = 3 \qquad [12.1k]$$

$$\gamma_k(y) = \cos\frac{(k-3)\pi y}{h} \quad \text{for } k \geq 4 \qquad [12.1l]$$

or in non-dimensional form as follows.

Non-dimensional set of admissible functions in the x direction

$$\hat{\phi}_i(\xi) = 1 \quad \text{for } i = 1 \qquad [12.2a]$$

$$\hat{\phi}_i(\xi) = \xi \quad \text{for } i = 2 \qquad [12.2b]$$

$$\hat{\phi}_i(\xi) = \xi^2 \quad \text{for } i = 3 \qquad [12.2c]$$

$$\hat{\phi}_i(\xi) = \cos(i-3)\pi\xi \quad \text{for } i \geq 4 \qquad [12.2d]$$

with $\xi = x/a$, and as follows.

Non-dimensional set of admissible functions in the y direction

$$\hat{\psi}_j(\eta) = 1 \quad \text{for } j = 1 \qquad [12.2e]$$

$$\hat{\psi}_j(\eta) = \eta \quad \text{for } j = 2 \qquad [12.2f]$$

$$\hat{\psi}_j(\eta) = \eta^2 \quad \text{for } j = 3 \qquad [12.2g]$$

$$\psi_j(\eta) = \cos(j-3)\pi\eta \quad \text{for } j \geq 4 \qquad [12.2h]$$

with $\eta = y/b$.

Non-dimensional set of admissible functions in the z direction

$$\hat{\gamma}_k(\zeta) = 1 \quad \text{for } k = 1 \qquad [12.2i]$$

$$\hat{\gamma}_k(\zeta) = \zeta \quad \text{for } k = 2 \qquad [12.2j]$$

$$\hat{\gamma}_k(\zeta) = \zeta^2 \quad \text{for } k = 3 \qquad [12.2k]$$

$$\hat{\gamma}_k(\zeta) = \cos(k-3)\pi\zeta \quad \text{for } k \geq 4 \qquad [12.2l]$$

with $\zeta = z/\mathrm{h}$.

The displacements of the 3D body in the x, y and z directions are given by:

$$u(x,y,z,t) = U(x,y,z)\sin\omega t \qquad [12.3a]$$

$$v(x,y,z,t) = V(x,y,z)\sin\omega t \qquad [12.3b]$$

$$w(x,y,z,t) = W(x,y,z)\sin\omega t \qquad [12.3c]$$

For convenience the same number of functions N will be used in all three directions x, y and z in the non-dimensional displacement amplitude functions $\hat{U}(\eta,\chi,\xi)$, $\hat{V}(\eta,\chi,\xi)$ and $\hat{W}(\eta,\chi,\xi)$ defined as:

$$\hat{U}(\xi,\eta,\zeta) = \sum_{i=1}^{N}\sum_{j=1}^{N}\sum_{k=1}^{N} A_{ijk}\hat{\phi}_i(\xi)\hat{\psi}_j(\eta)\hat{\gamma}_j(\zeta) \qquad [12.4a]$$

$$\hat{V}(\xi,\eta,\zeta) = \sum_{l=1}^{N}\sum_{m=1}^{N}\sum_{n=1}^{N} B_{lmn}\hat{\phi}_l(\xi)\hat{\psi}_m(\eta)\hat{\gamma}_n(\zeta) \qquad [12.4b]$$

$$\hat{W}(\xi,\eta,\zeta) = \sum_{p=1}^{N}\sum_{q=1}^{N}\sum_{r=1}^{N} C_{pqr}\hat{\phi}_p(\xi)\hat{\psi}_q(\eta)\hat{\gamma}_r(\zeta) \qquad [12.4c]$$

Then as in the previous chapters the energy terms need to be defined. The kinetic energy of the 3D body is [NAG 08]:

$$T = \frac{\rho}{2}\int_0^a\int_0^b\int_0^h\left(\left(\frac{\partial u}{\partial t}\right)^2 + \left(\frac{\partial v}{\partial t}\right)^2 + \left(\frac{\partial w}{\partial t}\right)^2\right) dz\, dy\, dx, \qquad [12.5a]$$

or in non-dimensional form:

$$\hat{T} = \frac{\rho\omega^2 a^3 bh}{2}\int_0^1\int_0^1\int_0^1\left(U^2 + V^2 + W^2\right) d\zeta\, d\eta\, d\xi \qquad [12.5b]$$

where ρ is the density of the material of the 3D body. The strain energy of the 3D body is [NAG 08]:

$$V_{\max} = \frac{1}{2}\int_0^a\int_0^b\int_0^h \left(\Delta_1 \Gamma_1 + 2\Delta_2 \Gamma_2 + G\Gamma_3\right)\, dz\, dy\, dx\,, \qquad [12.6a]$$

where $\Gamma_1 = \varepsilon_x^2 + \varepsilon_y^2 + \varepsilon_z^2$, $\Gamma_2 = \varepsilon_x\varepsilon_y + \varepsilon_y\varepsilon_z + \varepsilon_z\varepsilon_x$, $\Gamma_3 = \gamma_{xy}^2 + \gamma_{yz}^2 + \gamma_{zx}^2$

$$\Delta_1 = \frac{E(1-\nu)}{(1+\nu)(1-2\nu)}, \quad \Delta_2 = \frac{\nu E}{(1+\nu)(1-2\nu)} \text{ and } G = \frac{E}{2(1+\nu)},$$

where ε_i, γ_{ij} are normal and shear strains, while E and G are the Young's modulus and the shear modulus.

The strain energy of the 3D body in non-dimensional form is [NAG 08]:

$$\hat{V} = \frac{abhE}{2}\int_0^1\int_0^1\int_0^1 \left(\bar{\Delta}_1\bar{\Pi}_1 + \bar{\Delta}_2\bar{\Pi}_2 + \bar{\Delta}_3\bar{\Pi}_3\right)\, d\zeta\, d\eta\, d\xi\,, \qquad [12.6b]$$

where

$$\bar{\Delta}_1 = \frac{1-\nu}{(1+\nu)(1-2\nu)}, \quad \Delta_2 = \frac{\nu}{(1+\nu)(1-2\nu)} \text{ and } \Delta_3 = \frac{1}{2(1+\nu)}$$

$$\bar{\Pi}_1 = \left(\frac{\partial U}{\partial \xi}\right)^2 + \left(\frac{a}{b}\right)^2\left(\frac{\partial V}{\partial \eta}\right)^2 + \left(\frac{a}{h}\right)^2\left(\frac{\partial W}{\partial \zeta}\right)^2,$$

$$\begin{aligned}\bar{\Pi}_2 = {} & \left(\frac{a}{b}\right)\left[\left(\frac{\partial U}{\partial \xi}\right)\left(\frac{\partial V}{\partial \eta}\right) + \left(\frac{\partial V}{\partial \eta}\right)\left(\frac{\partial U}{\partial \xi}\right)\right] + \\ & \left(\frac{a}{b}\right)\left(\frac{a}{h}\right)\left[\left(\frac{\partial V}{\partial \eta}\right)\left(\frac{\partial W}{\partial \zeta}\right) + \left(\frac{\partial W}{\partial \zeta}\right)\left(\frac{\partial V}{\partial \eta}\right)\right] \\ & + \left(\frac{a}{h}\right)\left[\left(\frac{\partial W}{\partial \zeta}\right)\left(\frac{\partial U}{\partial \xi}\right) + \left(\frac{\partial U}{\partial \xi}\right)\left(\frac{\partial W}{\partial \zeta}\right)\right]\end{aligned}, \text{ and}$$

$$\bar{\Pi}_3 = \left(\frac{a}{b}\right)^2\left(\frac{\partial V}{\partial \eta}\right) + \left(\frac{a}{b}\right)\left[\begin{array}{l}\left(\frac{\partial U}{\partial \eta}\right)\left(\frac{\partial V}{\partial \xi}\right) + \\ \left(\frac{\partial V}{\partial \xi}\right)\left(\frac{\partial U}{\partial \eta}\right)\end{array}\right] + \frac{1}{2}$$

$$\left(\frac{\partial V}{\partial \xi}\right)^2 + \left(\frac{a}{h}\right)^2\left(\frac{\partial V}{\partial \zeta}\right) +$$

$$\left(\frac{a}{b}\right)\left(\frac{a}{h}\right)\left[\left(\frac{\partial V}{\partial \zeta}\right)\left(\frac{\partial W}{\partial \eta}\right) + \left(\frac{\partial W}{\partial \eta}\right)\left(\frac{\partial V}{\partial \zeta}\right)\right]$$

$$+\left(\frac{a}{b}\right)^2\left(\frac{\partial W}{\partial \eta}\right)^2 + \left(\frac{\partial W}{\partial \xi}\right)^2 +$$

$$\left(\frac{a}{h}\right)\left[\left(\frac{\partial W}{\partial \xi}\right)\left(\frac{\partial U}{\partial \zeta}\right) + \left(\frac{\partial U}{\partial \zeta}\right)\left(\frac{\partial W}{\partial \xi}\right)\right]$$

$$+\left(\frac{a}{h}\right)^2\left(\frac{\partial U}{\partial \zeta}\right)^2$$

Carrying out the Rayleigh–Ritz minimization of the non-dimensional kinetic energy gives the terms of the mass matrix $M^{uu}_{ijk\bar{i}\bar{j}\bar{k}}$, $M^{vv}_{lmn\bar{l}\bar{m}\bar{n}}$ and $K^{ww}_{pqr\bar{p}\bar{q}\bar{r}}$ as follows:

$$M^{uu}_{ijk\bar{i}\bar{j}\bar{k}} = (1-\nu)\,E^{(0,0)}_{ui u\bar{i}}\,F^{(0,0)}_{uj u\bar{j}}\,G^{(0,0)}_{uk u\bar{k}} \qquad [12.7a]$$

$$M^{vv}_{lmn\bar{l}\bar{m}\bar{n}} = (1-\nu)\,E^{(0,0)}_{vl v\bar{l}}\,F^{(0,0)}_{vm v\bar{m}}\,G^{(0,0)}_{vn v\bar{n}} \qquad [12.7b]$$

$$M^{ww}_{pqr\bar{p}\bar{q}\bar{r}} = (1-\nu)\,E^{(0,0)}_{wp w\bar{p}}\,F^{(0,0)}_{wq w\bar{q}}\,G^{(0,0)}_{wr w\bar{r}} \qquad [12.7c]$$

These six-dimensional arrays can be written as a square matrix of size N^3. Thus the size of the square and symmetric mass matrix $\mathbf{M}$ is equal to $3N^3$ and is built with the terms described in equation [12.7] as shown below.

$$\mathbf{M} = \begin{bmatrix} M^{uu}_{ijk\bar{i}\bar{j}\bar{k}} & 0 & 0 \\ 0 & M^{vv}_{lmn\bar{l}\bar{m}\bar{n}} & 0 \\ 0 & 0 & M^{ww}_{pqr\bar{p}\bar{q}\bar{r}} \end{bmatrix} \qquad [12.8]$$

The minimization of the strain energy of the 3D body V_{max} with respect to the unknown coefficients A_{ijk}, B_{lmn} and C_{pqr} gives the terms of the stiffness matrix defined as follows:

$$K^{uu}_{ijk\bar{i}\bar{j}\bar{k}} = \frac{1-\nu}{1-2\nu} E^{(1,1)}_{ui u\bar{i}} F^{(0,0)}_{uj u\bar{j}} G^{(0,0)}_{uk u\bar{k}} + \frac{1}{2}\left(\frac{a}{b}\right)^2 E^{(0,0)}_{ui u\bar{i}} F^{(1,1)}_{uj u\bar{j}} G^{(0,0)}_{uk u\bar{k}} + \\ + \frac{1}{2}\left(\frac{a}{h}\right)^2 E^{(0,0)}_{ui u\bar{i}} F^{(0,0)}_{uj u\bar{j}} G^{(1,1)}_{uk u\bar{k}} \tag{12.9a}$$

$$K^{uv}_{ijklmn} = \left(\frac{a}{b}\right)\left(\frac{\nu}{1-2\nu} E^{(1,0)}_{uivl} F^{(0,1)}_{ujvm} G^{(0,0)}_{ukvn} + \frac{1}{2} E^{(0,1)}_{uivl} F^{(1,0)}_{ujvm} G^{(0,0)}_{ukvn}\right) \tag{12.9b}$$

$$K^{uw}_{ijkpqr} = \left(\frac{a}{h}\right)\left(\frac{\nu}{1-2\nu} E^{(1,0)}_{uivp} F^{(0,0)}_{ujvq} G^{(0,1)}_{ukvr} + \frac{1}{2} E^{(0,1)}_{uivp} F^{(0,0)}_{ujvq} G^{(1,0)}_{ukvr}\right) \tag{12.9c}$$

$$K^{vv}_{lmn\bar{l}\bar{m}\bar{n}} = \left(\frac{a}{b}\right)^2 \frac{1-\nu}{1-2\nu} E^{(0,0)}_{vl v\bar{l}} F^{(1,1)}_{vm v\bar{m}} G^{(0,0)}_{vn v\bar{n}} + \frac{1}{2}\left(\frac{a}{h}\right)^2 E^{(0,0)}_{vl v\bar{l}} F^{(0,0)}_{vm v\bar{m}} G^{(1,1)}_{vn v\bar{n}} + \\ \frac{1}{2} E^{(1,1)}_{vl v\bar{l}} F^{(0,0)}_{vm v\bar{m}} G^{(0,0)}_{vn v\bar{n}} \tag{12.9d}$$

$$K^{vw}_{lmnpqr} = \left(\frac{a^2}{bh}\right)\left(\frac{\nu}{1-2\nu} E^{(0,0)}_{vlwp} F^{(1,0)}_{vmwq} G^{(0,1)}_{vnwr} + \frac{1}{2} E^{(0,0)}_{vlwp} F^{(0,1)}_{vmwq} G^{(1,0)}_{vnwr}\right) \tag{12.9e}$$

$$K^{ww}_{pqr\bar{p}\bar{q}\bar{r}} = \left(\frac{a}{h}\right)^2 \left(\frac{1-\nu}{1-2\nu} E^{(0,0)}_{wp w\bar{p}} F^{(0,0)}_{wq w\bar{q}} G^{(1,1)}_{wr w\bar{r}}\right) + \frac{1}{2}\left(\frac{a}{b}\right)^2 E^{(0,0)}_{wp w\bar{p}} F^{(1,1)}_{wq w\bar{q}} G^{(0,0)}_{wr w\bar{r}} + \\ + \frac{1}{2}\left(\frac{a}{b}\right)^2 E^{(1,1)}_{wp w\bar{p}} F^{(0,0)}_{wq w\bar{q}} G^{(0,0)}_{wr w\bar{r}} \tag{12.9f}$$

Again the terms defined above are six-dimensional arrays that can be written as a square matrix of size N^3. Thus the size of the square and symmetric stiffness matrix $\mathbf{K}$ is equal to $3N^3$ and is built with the terms described in [12.9] as shown below.

$$\mathbf{K} = \begin{bmatrix} K^{uu}_{ijk\bar{i}\bar{j}\bar{k}} & K^{uv}_{ijklmn} & K^{uw}_{ijkpqr} \\ & K^{vv}_{lmn\bar{l}\bar{m}\bar{n}} & K^{vw}_{lmnpqr} \\ Symm & & K^{ww}_{pqr\bar{p}\bar{q}\bar{r}} \end{bmatrix} \tag{12.10}$$

Adding constraints

Considering the stiffness coefficient of all springs to have the same value k, the strain energy of the artificial springs along the sides of the 3D body perpendicular to plane xy in a similar way as in [NAG 08] is:

$$V_{spring} = \frac{k}{2}\left(\begin{array}{l} \int_0^h \int_0^b \left(u^2 + v^2 + w^2\right) dy\, dz \Big|_{x=0,a} + \\ \int_0^h \int_0^a \left(u^2 + v^2 + w^2\right) dx\, dz \Big|_{y=0,b} \end{array} \right) \qquad [12.11a]$$

Or in non-dimensional form:

$$\hat{V}_{spring} = \frac{\hat{p}abhE}{2}\left(\begin{array}{l} \int_0^1 \int_0^1 \left(U^2 + V^2 + W^2\right) d\eta\, d\zeta \Big|_{\xi=0,1} + \\ \frac{a}{b} \int_0^1 \int_0^1 \left(U^2 + V^2 + W^2\right) d\xi\, d\zeta \Big|_{\eta=0,1} \end{array} \right), \qquad [12.11b]$$

where $\hat{p} = ka/E$ is the non-dimensional penalty parameter representing the stiffness of artificial springs used to model all constraints. For instance, to constrain the side at $x = 0$ in the x direction, the following terms must be computed.

$$P^{uu}_{ijk\bar{i}\bar{j}\bar{k}} = \hat{p}\left[\hat{\varphi}_i(0)\hat{\varphi}_{\bar{i}}(0) F^{(0,0)}_{j\bar{j}} G^{(0,0)}_{k\bar{k}}\right] \qquad [12.12]$$

And the penalty matrix for the 3D body in cantilever can be built as:

$$\mathbf{P} = \begin{bmatrix} P^{uu}_{ijk\bar{i}\bar{j}\bar{k}} & 0 & 0 \\ 0 & P^{vv}_{lmn\bar{l}\bar{m}\bar{n}} & 0 \\ 0 & 0 & P^{ww}_{pqr\bar{p}\bar{q}\bar{r}} \end{bmatrix}, \qquad [12.13]$$

where the three submatrices are identical if the same number of terms are used in all directions.

Alternatively, penalty terms can be added to the mass matrix as shown by [ILA 05] in a similar way as in previous chapters. If artificial mass is used to model constraints, the penalty matrix has to be added to the mass matrix of the system.

Applying conservation of energy and the Rayleigh–Ritz minimization as in previous chapters leads to the following eigenvalue equation:

$$\left(\mathbf{K}-\lambda^2\mathbf{M}\right)\mathbf{C}=\mathbf{0}, \qquad [12.14]$$

where $\mathbf{0}$ is a null vector, $\mathbf{C}$ are column vectors of unknown coefficients and λ are the non-dimensional frequency parameters. Results for unconstrained 3D bodies obtained with the present method are compared with those reported in [NAG 08] where λ is defined as:

$$\lambda=\omega b^2\sqrt{\rho h/D} \qquad [12.15a]$$

All other results are compared with those presented in [LIE 93] who defined λ as:

$$\lambda=\frac{\omega b^2}{\pi^2}\sqrt{\rho h/D}, \qquad [12.15b]$$

where D is the flexural rigidity of the plates $D=Eh^3/12(1-\upsilon^2)$.

All submatrices needed to solve 3D bodies were already computed in the previous chapters to solve for natural frequencies of beams, plates and shells.

12.2. Results

The first set of results for non-dimensional frequency parameters λ of 3D bodies presented in Table 12.1 corresponds to a body with the following geometrical properties: square base ($a/b=1$) and height to side ratio $h/a=0.5$. Table 12.1 presents the convergence results with respect to the number of functions N included in the Rayleigh–Ritz procedure for an unconstrained 3D body. In all cases in this work, the value of the Poisson's ratio ν was fixed to 0.3. Results on Table 12.1 are compared with those given by Nagino *et al* [NAG 08] and with results obtained using the commercial finite element code Abaqus.

Results presented in [NAG 08] in Table 12.1 were obtained using 13 × 13 × 9 terms in the sets of admissible functions, while Abaqus results were obtained using 500 8-node linear brick elements (C3D8). Table 12.1 shows that converged results were obtained to 3 decimal places with $\overline{N}=11$. Monotonic convergence from above as expected when using the RRM is observed.

	Frequency parameters $\lambda = \omega b^2 \sqrt{\rho h / D}$					
N	1	2	3	4	5	6
4	8.835	13.342	16.299	16.658	18.029	18.029
5	8.822	12.548	15.036	16.117	17.062	17.062
6	8.787	12.523	14.973	16.080	17.041	17.041
7	8.786	12.518	14.967	16.075	17.033	17.033
8	8.782	12.516	14.964	16.074	17.032	17.032
9	8.781	12.515	14.963	16.073	17.031	17.031
10	8.781	12.515	14.962	16.073	17.030	17.030
11	8.780	12.515	14.962	16.073	17.030	17.030
12	8.780	12.515	14.962	16.073	17.030	17.030
[NAG 08]	8.780	12.515	14.961	16.072	17.030	17.030
Abaqus	8.961	12.624	15.004	16.608	17.275	17.275

Table 12.1. *Frequency parameters of a rectangular three-dimensional element in unconstrained condition with height to side ratio* $h / a = 0.5$

All the following results were compared with those presented in [LIE 93]. For the cases presented here, results in [LIE 93] were obtained using $6 \times 6 \times 8$ terms in the x, y and z directions using four symmetry classes which helps in reducing the size of the matrices of the eigenproblems. The four different types of symmetry on the xz and yz planes are double symmetry modes (SS), symmetric–antisymmetric modes (SA), antisymmetric–symmetric modes (AS) and double antisymmetric modes (AA). Tables 12.2–12.4 show the type of symmetry of the corresponding results. Table 12.2 presents results for a 3D body in cantilever. Table 12.3 presents results for a 3D body with boundary conditions SFSF on four sides. This case has one rigid body mode for which the frequency parameter is equal to zero. Table 12.3 presents results for a 3D body with boundary conditions CCCC on four sides.

		Frequency parameters $\lambda = \frac{\omega b^2}{\pi^2} \sqrt{\rho h / D}$					
N	$\hat{p}$	1 SA	2 AS	3 AA	4 SS	5 SA	6 AS
12	10^6	0.2976	0.4450	0.5263	1.0661	1.1088	1.1864
[LIE 93]		0.2976	0.4449	0.5263	1.0687	1.1087	1.1863

Table 12.2. *Frequency parameters of a rectangular three-dimensional element in cantilever with height to side ratio* $h / a = 0.5$

		Frequency parameters $\lambda = \frac{\omega b^2}{\pi^2}\sqrt{\rho h / D}$					
N	$\hat{p}$	1 SS	2 AS-1	3 AS-2	4 AA-1	5 SA-2	6 SA-3
12	10^5	0.7261	0.9874	1.0223	1.3047	1.8451	1.8503
[LIE 93]		0.7261	0.9874	1.0223	1.3047	1.8451	1.8502

Table 12.3. *Frequency parameters of a rectangular three-dimensional element with boundary conditions SFSF on four sides with height to side ratio* $h / a = 0.5$

		Frequency parameters $\lambda = \frac{\omega b^2}{\pi^2}\sqrt{\rho h / D}$					
N	$\hat{p}$	1 SS	2 AS-1	3 SA-1	4 AS-2	5 SA-2	6 AA-1
12	10^5	1.5509	2.4406	2.4406	2.5161	2.5161	2.9766
[LIE 93]		1.5496	2.4396	2.4396	2.5151	2.5151	2.9765

Table 12.4. *Frequency parameters of a rectangular three-dimensional element with boundary conditions CCCC on four sides with height to side ratio* $h / a = 0.5$

13

Vibration of Axially Loaded Beams and Geometric Stiffness

13.1. Introduction

So far we have seen the application of the Rayleigh–Ritz method to solve one particular type of eigenvalue problem, namely determining the natural frequencies and modes. Another type of eigenvalue problem encountered by structural engineers is the determination of elastic critical loads. There is also an interesting relationship between the natural frequencies and critical loads (also commonly referred to as buckling loads), which is that the natural frequencies tend to zero as a structure approaches a critical state. This will be discussed as a part of this chapter which deals with the derivation of the potential energy term associated with a static axial force when considering the lateral vibration of a beam; although this is not the main meal bear in mind that a side dish could make a meal more enjoyable. Just as the elastic potential energy gives rise to the elastic stiffness matrix, the axial force effect gives rise to what is called the geometric stiffness matrix.

In the free vibration analysis of beams, the derivation of the total potential energy of a statically axially loaded straight beam undergoing small amplitude vibration often causes confusion. While the potential energy associated with the dynamic bending is easily understood, the contribution of the axial load to the total potential energy is often not clearly understood. The additional potential energy due to the axial load is usually obtained by multiplying the end (axial) load by the relative dynamic displacement of the two ends [MCC 73]. This derivation is valid only if at least one of the ends were free to move longitudinally during vibration. However, the same expression for the dynamic component of the total potential energy is also applicable if the two ends were longitudinally restrained during vibration (see [SOU 69]). This would be the case if the axial force were induced by a change in temperature in a beam with longitudinally restrained ends. It is well known that the solution to the differential equation governing the small amplitude vibration of a statically axially loaded thin beam is independent of longitudinal boundary conditions [TIM 55]. We will see that the same total potential energy expression is applicable for any longitudinal end conditions, and is therefore applicable for beams with practical longitudinal end conditions (which may be neither fully free, nor fully restrained). The effect of partial lateral restraints is included for the sake of completeness, and its application to a cable stayed beam/column is described. This approach may also be of some use in the dynamic analysis of robot manipulators with partial restraints that provide flexible joints [WAR 94].

In the present derivations, the induced static axial force is treated as a function of the longitudinal co-ordinate x, to allow for the inclusion of the possible effects of body force due to factors such as centrifugal acceleration or gravity. While a large number of publications [YOK 88] and [SCH 85] are available on the use of energy principles to solve such problems that have many important practical applications (for example dynamic analysis of rotating beams and slender columns), the derivation of the potential energy terms appears to have received little attention. Where presented, the energy formulations are derived from the equations of motion using variational principles [SCH 85].

13.2. The potential energy due to a static axial force in a vibrating beam

Consider the small amplitude lateral vibration of a statically axially loaded straight Euler–Bernoulli beam having length L (at the statically loaded equilibrium state), cross-sectional area A, Young's modulus E and second moment of area about the neutral axis of bending I. It is subject to a static body force of intensity $q(\mathrm{x})$ per unit length, and static axial end loads (or reactions) $P(0)$ and $P(L)$ as shown in Figure 13.1. The resulting static axial force $P(x)$ is taken as being compressive. The result for a tensile axial force $T(x)$ is obtained simply by replacing $P(x)$ with $-T(x)$. No specific boundary conditions are assumed in the derivations, though the figure corresponds to that of simply supported conditions. The possibility of longitudinal displacements at the ends is permitted in the diagram, but the derivations are valid for fully longitudinally restrained end conditions too, as it is only a special case of longitudinal displacements being zero.

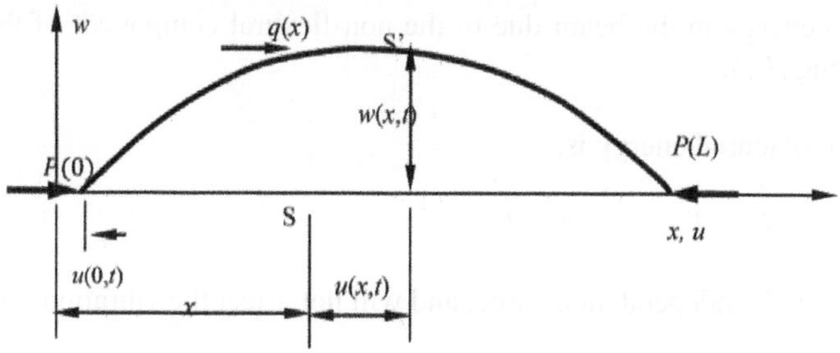

Figure 13.1. *Deflected shape of the vibrating beam*

In the figure, a point on the centroidal axis of the beam at a distance x from the left-hand end (origin) when the beam is in static equilibrium under the axial force is labeled S. The position of the same point, at time t during vibration is labeled S′. $w(x,t)$ denotes the lateral displacement of the beam from the straight, static equilibrium position, and $u(x,t)$ is the longitudinal displacement.

In this paper, the term "longitudinal" is used for the x direction (along the axis of the undeflected beam), while "axial" is used to describe the direction of the centroidal axis of the beam (a time dependent direction). "Lateral" direction is perpendicular to the x axis and is along w, while 'transverse' direction (z) is perpendicular to the axis of the beam (time dependent).

In addition to the standard assumptions for the static analysis of an Euler–Bernoulli beam, it is assumed that the vibration is primarily lateral, and that the amplitude of lateral displacement is sufficiently small so that any longitudinal motion is secondary. This means that if $w, \partial w/\partial x$ and $\partial^2 w/\partial x^2$ are first order dynamic terms, u and $\partial u/\partial x$ would be second order terms.

The total potential energy of the beam (V) consists of the following:

1) potential energy of the beam when it is in static equilibrium (Vs);

2) strain energy in the beam due to the dynamic bending (Ub);

3) change in potential energy due to the longitudinal dynamic displacement of the end loads (or reactions) (V_L);

4) change in potential energy due to the lateral dynamic translation and dynamic rotation of the support forces and support moments (or reactions) (V_R);

5) change in potential energy due to the work done by the body forces during vibration (V_q);

6) strain energy in the beam due to the non-flexural component of the dynamic axial straining (U_a);

The total potential energy is:

$$V = V_s + U_b + V_L + V_R + V_q + U_a \qquad [13.1]$$

Of these, V_s is independent of time, and will not affect the vibration analysis.

The dynamic bending strain at distance y from the centroidal axis is given by:

$$\varepsilon = -y\partial^2 w / \partial x^2$$

The corresponding strain energy (given in many standard books) is:

$$U_b = \int_0^L \frac{EI}{2}\left(\frac{\partial^2 w}{\partial x^2}\right)^2 dx \qquad [13.2]$$

This being a square of a small first order term is a second order term. The change in potential energy due to the longitudinal dynamic displacement of the end loads is given by:

$$V_L = -P(0)\,u(0,t) + P(L)\,u(L,t) \qquad [13.3]$$

This is also a second order term (product of a constant static term and a second order dynamic term). A fluctuation in the end forces is possible for beams with longitudinally restrained ends, but its contribution to V would be a fourth order term which is negligible for small amplitude vibrations.

It should be noted that $u(x,t)$ is measured in the positive x direction and that the potential energy is positive if the dynamic displacement u is against the load P.

The potential energy due to the lateral translation and rotation of the support forces (V_R) would be non-zero only if the boundaries were partially restrained, in which case it would depend on the support stiffnesses. It is worth mentioning that, partial lateral restraints may be effected by pre-tensioned cables even if they were connected longitudinally to the beam. For any combination of the standard boundary conditions (that is: free, simply supported, rotationally restrained laterally free or clamped) V_R would be zero. The longitudinal boundary conditions have no effect on V_R.

The change in potential energy due to the work done by the body force during vibration is:

$$V_q = -\int_0^L qu \, \mathrm{d}x \qquad [13.4]$$

For longitudinal equilibrium, $dP / dx = q$,

Substituting this into equation [13.4] gives the following second order term:

$$V_q = -\int_0^L (dP / \mathrm{d}x) u \, \mathrm{d}x \qquad [13.5]$$

As a result of vibration, in addition to the bending strain distribution, an axial strain distribution would also be induced. While the bending strain varies linearly at any given transverse section, this additional axial strain remains constant, and is smaller in magnitude compared to the maximum bending strain for small amplitude lateral vibrations. This strain is due to the longitudinal and lateral motions of the beam, and will be referred to *as "the non-flexural dynamic axial strain*" in this paper, and is denoted by ε'. The corresponding "*non-flexural dynamic axial stres*s" is denoted by σ'. Let the static axial stress and strain be σ_s and ε_s respectively. Then the net non-flexural stress and strain at time t is given by:

$$\sigma = \sigma_s + \sigma' \qquad [13.6a]$$

$$\varepsilon = \varepsilon_s + \varepsilon' \qquad [13.6b]$$

where $\sigma_s = - P/A$, $\varepsilon_s = -P / (AE)$ and $\sigma' = E\varepsilon'$ [13.6c,d,e]

Let us now consider the non-flexural axial straining of an infinitesimal element of the beam at distance x from the origin as shown in Figure 13.2.

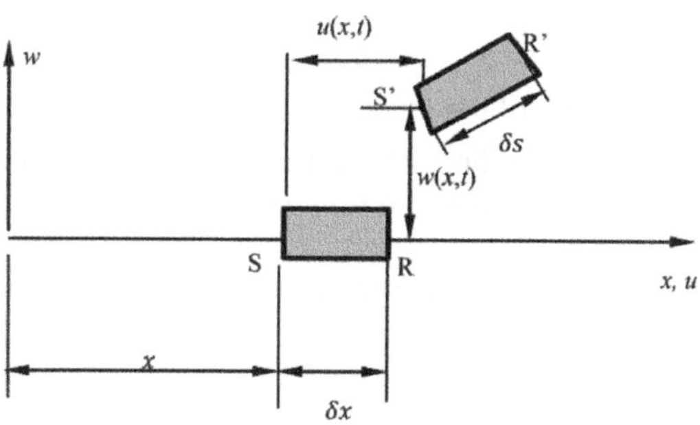

Figure 13.2. *The dynamic straining and displacement of an infinitesimal element*

The axial length of the element when it is in static equilibrium under the given axial loading is δx. Let the corresponding length at the centroidal axis at time t during vibration be δs.

$$\delta s = \sqrt{(\delta x + \delta u)^2 + \delta w^2} \qquad [13.7a]$$

The non-flexural dynamic axial strain ε' is obtained by taking the dynamic axial strain at the centroidal axis which is given by:

$$\varepsilon' = \text{Lim}_{\delta x \to 0} \frac{(\delta s - \delta x)}{\delta x}$$

Substituting equation [13.7a] into the above equation gives:

$$\varepsilon' = \left((1 + \partial u / \partial x)^2 + (\partial w / \partial x)^2\right)^{1/2} - 1$$

Using the binomial expansion and neglecting the terms that are smaller than second order gives:

$$\varepsilon' = \varepsilon'_u + \varepsilon'_w \qquad [13.7b]$$

where ε'_w is the non-flexural dynamic axial strain due to lateral motion and is given by:

$$\varepsilon'_w = (1/2)(\partial w / \partial x)^2 \qquad [13.8a]$$

and ε'_u is the non-flexural dynamic axial strain due to longitudinal motion and is given by:

$$\varepsilon'_u = \partial u / \partial x \qquad [13.8b]$$

Although ε'_w and ε'_u are of the second order, and are small compared to the first order term for the dynamic bending strain, their contribution to the potential energy is not negligible as can be seen from the following derivation.

The strain energy of the infinitesimal element (δx) due to the static axial straining is given by:

$$\delta U_{a,s} = \frac{A}{2}(\sigma_s \varepsilon_s)\delta x$$

The strain energy of the same element due to non-flexural axial straining at time t is:

$$\delta U_a = \frac{A}{2}(\sigma\varepsilon)\delta x$$

Hence the strain energy due to the non-flexural dynamic axial straining is:

$$\delta U'_a = \delta U_a - \delta U_{a,s} = \frac{A}{2}(\sigma\varepsilon - \sigma_s\varepsilon_s)\delta x$$

Substituting equations [13.6a] and [13.6b] into the above equation gives:

$$\delta U'_a = \frac{A}{2}(\sigma_s\varepsilon' + \sigma'\varepsilon_s + \sigma'\varepsilon')\delta x$$

Neglecting the last term in the above equation which is a fourth order term (product of two second order terms) gives:

$$\delta U'_a = \frac{A}{2}(\sigma_s\varepsilon' + \sigma'\varepsilon_s)\delta x$$

Substituting equations [13.6c]–[13.6e] into the above gives:

$$\delta U'_a = -P\varepsilon'\delta x$$

Hence,

$$U'_a = -\int_0^L P\varepsilon' dx \qquad [13.9]$$

Substituting equation [13.7b] into equation [13.9] gives:

$$U'_a = -\int_0^L P(\varepsilon'_u + \varepsilon'_w)dx$$

Using equations [13.8a] and [13.8b] this becomes

$$U'_a = -\int_0^L P\left(\frac{\partial u}{\partial x} + \frac{1}{2}\left(\frac{\partial w}{\partial x}\right)^2\right)dx = [Pu]_0^L - \int_0^L\left(u\frac{\partial P}{\partial x} + \frac{P}{2}\left(\frac{\partial w}{\partial x}\right)^2\right)dx$$

This becomes:

$$U'_a = -P(L)u(L,t) + P(0)u(0,t) - \int_0^L \left(u\frac{\partial P}{\partial x} + \frac{P}{2}\left(\frac{\partial w}{\partial x}\right)^2 \right) dx \qquad [13.10]$$

All terms in the above equation are of second order, resulting in a non-flexural strain energy comparable to the strain energy due to bending. Substituting equations [13.2], [13.3], [13.5] and [13.10] into equation [13.1] gives the required total potential energy V,

$$\text{i.e. } V = \int_0^L \frac{EI}{2}\left(\frac{\partial^2 w}{\partial x^2}\right)^2 dx - \int_0^L \frac{P}{2}\left(\frac{\partial w}{\partial x}\right)^2 dx + V_R + V_s \qquad [13.11]$$

The terms associated with the longitudinal motion u vanish. The total potential energy is therefore independent of longitudinal boundary conditions during vibration. While the body force term does not appear explicitly in equation [13.11], it has an indirect effect through the induced static axial force P. Equation [13.11] agrees with the total potential energy expressions given in the literature [MCC 73, YOK 88, SCH 85].

13.3. Determination of natural frequencies

The total potential energy expression obtained results in an expression for the fundamental natural frequency using Rayleigh's principle. The following analysis shows that the result is independent of longitudinal end conditions. Let ρ be the density of the beam, and

$$w = \varphi(x).\sin(\omega t + \alpha) \qquad [13.12]$$

where φ is a permissible shape function which satisfies the geometrical boundary conditions for the beam. For simplicity let us assume that the lateral boundary conditions correspond to a combination of the standard cases. This means that V would be zero. The kinetic energy of the beam (neglecting longitudinal inertia) is:

$$\frac{1}{2}\int_0^L \rho A\left(\frac{\partial w}{\partial t}\right)^2 dx \qquad [13.13]$$

From equations [13.12] and [13.13] the maximum kinetic energy of the beam is obtained as:

$$T_m = \frac{\omega^2}{2}\int_0^L \rho A \varphi^2 dx \qquad [13.14]$$

The maximum total potential energy during vibration is obtained by setting $w = \phi$ in equation [13.11] which gives:

$$V = \int_0^L \frac{EI}{2}\left(\frac{\partial^2 \varphi}{\partial x^2}\right)^2 dx - \int_0^L \frac{P}{2}\left(\frac{\partial \varphi}{\partial x}\right)^2 dx + V_s + V_R$$

Using primes to denote differentiation w.r.t. x,

$$V = \int_0^L \frac{EI}{2}\varphi''^2 dx - \int_0^L \frac{P}{2}\varphi'^2 dx + V_s + V_R \qquad [13.15]$$

Using Rayleigh's principle $T_m = V_m$ gives an upper bound for the fundamental natural frequency. Substituting equations [13.14] and [13.15] into this equation results in the following expression:

$$\omega^2 = \frac{\int_0^L EI\varphi''^2 dx - \int_0^L P\varphi'^2 dx}{\int_0^L \rho A \varphi^2 dx} \qquad [13.16]$$

This agrees with the result in article 518 of reference [SOU 69], in which a tensile force T appears in place of $-P$. The right hand side of equation [13.16], the Rayleigh quotient, gives an upper bound for the fundamental natural frequency of flexural vibration of the beam, and for a uniform simply supported beam subject to a constant static axial compressive force P, using sine $(\pi x/L)$ for $\varphi(x)$ gives:

$$\omega^2 = (\pi^4 EI / L^4 - \pi^2 P / L^{2)} / (\rho A) \qquad [13.17]$$

As the function used is the first natural mode of vibration, this is the exact expression for the fundamental natural frequency. Generally the calculated frequency would be an upper bound to the exact fundamental natural frequency, and in Rayleigh–Ritz method φ is taken as the sum of a series of the products of permissible functions and undetermined coefficients, and the Rayleigh quotient is minimized with respect to the undetermined coefficients to obtain the lowest of the upper bound frequencies. Since equation [13.17] and the choice for ϕ do not depend on the longitudinal end conditions the calculated frequencies are independent of these conditions.

From the preceding derivations it is clear that the total potential energy and the natural frequencies are independent of longitudinal boundary conditions.

13.3.1. *The effect of partial lateral restraints*

For beams that are supported on partial lateral restraints, the potential energy changes as a result of the lateral translation (or rotation) of the support forces (or moments). As the displacements are against the support actions, the result is a gain in potential energy which is denoted as V_R in equation [13.12]. This energy is equal to the potential energy stored in the restraints and may be obtained as shown in the following example.

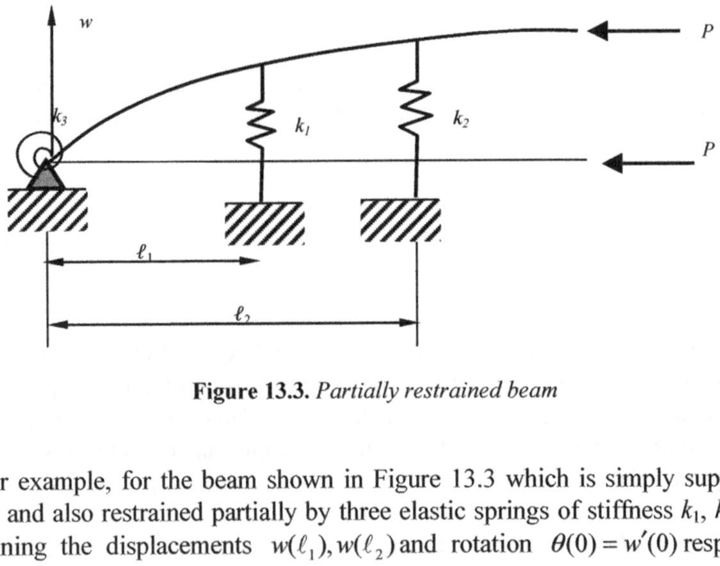

Figure 13.3. *Partially restrained beam*

For example, for the beam shown in Figure 13.3 which is simply supported at x = 0, and also restrained partially by three elastic springs of stiffness k_1, k_2, and k_3 restraining the displacements $w(\ell_1), w(\ell_2)$ and rotation $\theta(0) = w'(0)$ respectively. The potential energy due to the restraints is:

$$V_R = (1/2)\left(k_1\left(w(\ell_1)\right)^2 + k_2\left(w(\ell_2)\right)^2 + k_3\left(w'(0)\right)^2\right)$$

Note that the permissible shape function(s) for w need to satisfy only the fully restrained condition that $w(0) = 0$.

The effect of partial restraining of pre-tensioned cables can also be considered in a similar manner as illustrated in the following example.

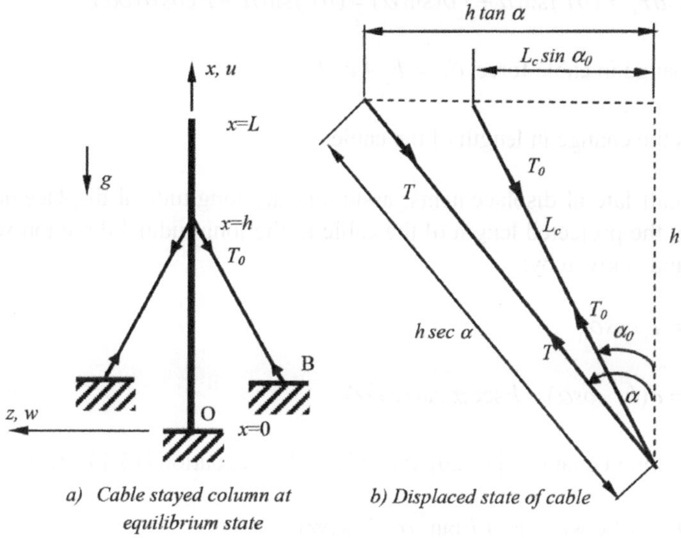

Figure 13.4. *Cable stayed column*

Consider the gravity loaded slender cantilever column in Figure 13.4(a) which is laterally partially restrained by two light cables of length L_c, cross-sectional area A_c and Young's modulus E_c. The properties of the column are denoted by the same symbols as those used for the beam in the preceding derivations (i.e. E, I, A and ρ). The gravitational acceleration is g. The cables are subject to a static tension T_0, and are connected to the column at distance l from the base. The angle between the cables and the axis of the column at the static equilibrium state is α_0. The potential energy associated with the lateral dynamic displacement of the cable forces at the connecting point at $x = h$ is given by:

$$V_R = (1/2)k_e\Delta^2 \qquad [13.18]$$

where Δ is the lateral dynamic displacement of the column at $x=h$ given by $\Delta = w(h)$, and k_e is the effective lateral stiffness of the two cables which can be determined as follows:

Figure 13.4(b) shows the displaced configuration of the cable on the right side. T is the force in the displaced cable. The lateral component of the cable force is given by:

$$F_\ell = T\sin\alpha$$

Hence $\delta F_\ell = (\delta T)\sin\alpha + T\delta(\sin\alpha) = (\delta T)\sin\alpha + T\cos\alpha(\delta\alpha)$ [13.19]

The change in cable force $\delta T = E_c A_c e / L_c$ [13.20]

where e is the change in length of the cable.

For small lateral displacements, assuming the longitudinal displacements to be negligible, the projected length of the cable in the longitudinal direction would be a constant and is given by:

$$h = L_c cos\alpha_0 \quad [13.21]$$

$$e = \delta(h / cos\alpha) = h\sec\alpha\tan\alpha(\delta\alpha) \quad [13.22]$$

Substituting equations [13.20] and [13.22] into equation [13.19] gives

$$\delta F_\ell = (T\cos\alpha + E_c A_c h\tan^2\alpha / L_c)(\delta\alpha) \quad [13.23]$$

Also $\Delta = \delta(h\tan\alpha) = h\sec^2\alpha(\delta\alpha)$

This gives $\delta\alpha = \Delta / (h\sec^2\alpha)$ [13.24]

Substituting equation [13.24] into equation [13.23] gives:

$$\delta F_\ell = (T\cos\alpha + E_c A_c h\tan^2\alpha / L_c)(\Delta / (h\sec^2\alpha)) \quad [13.25]$$

The lateral force required to produce a small displacement Δ from the static equilibrium state can be obtained by substituting α_0 for α, and T_0 for T together with equation [13.23] into equation [13.25]. This yields

$$\delta F_\ell / \Delta = (T_0\cos^2\alpha_0 + E_c A_c\sin^2\alpha_0) / L_c$$

By definition this is the effective lateral stiffness due to the left hand side cable. Both cables result in a combined stiffness given by:

$$k_e = 2(T_0\cos^2\alpha_0 + E_c A_c\sin^2\alpha_0) / L_c$$

Using the above in equation [13.18], the required potential energy due to the partial restraining of the cables is obtained.

$$V_R = (T_0 \cos^2 \alpha_0 + E_c A_c \sin^2 \alpha_0)\Delta^2 / L_c \qquad [13.26]$$

In this example the axial force P has a discontinuity at h and the resulting total potential energy can be shown to be

$$V = \int_0^L \frac{EI}{2} w''^2 dx - \int_0^h \frac{2T_0 \cos\alpha_0}{2} w'^2 dx - \int_0^L \frac{\rho g A(L-x)}{2} w'^2 dx + V_s + V_R$$

where V_R is given by equation [13.26].

13.3.2. *Summary*

The total potential energy of an initially straight, statically axially loaded Euler–Bernoulli beam undergoing small amplitude lateral vibrations is independent of longitudinal boundary conditions during vibration and is given by:

$$V = \int_0^L \frac{EI}{2}\left(\frac{\partial^2 w}{\partial x^2}\right)^2 dx - \int_0^L \frac{P}{2}\left(\frac{\partial w}{\partial x}\right)^2 dx + V_s + V_R$$

where P is the compressive static axial force induced in the beam. This formula is also applicable for non-uniform beams, and for cases where the axial force P varies as a result of body forces.

13.3.3. *Limitations of the above derivations*

The above formula is applicable only for initially straight Euler–Bernoulli beams undergoing small amplitude lateral vibrations. It has been shown [ILA 90] that the natural frequencies of slightly curved beams are dependent on longitudinal end conditions, even for small amplitude vibrations. The initial curvature would result in an extra non-flexural dynamic axial strain [ILA 86] which is of the first order and is sensitive to longitudinal end conditions. This chapter has now gone and replaced by a chapter on FEA.

For moderate or large amplitude vibrations the strain energy corresponding to the work done by the dynamic non-flexural stress over the corresponding strain should be included in the derivations which would lead to an amplitude dependent solution for the natural frequencies (for example see [THO 88]). The strain energy associated with bending may also require modification if large curvatures are caused during vibration. The kinetic energy due to the longitudinal motion (neglected in this work) may have to be included for higher modes, and for beams having moderate initial curvature.

13.4. Natural frequencies and critical loads of an Euler–Bernoulli beam

To illustrate the relationship between the natural frequencies and critical loads, let us consider the vibration of a simply supported beam as it is easy to obtain exact solutions.

$$w = f(x)\sin(\omega t+\alpha) = \sum_{i=1}^{n} G_i \varphi_i(x)\sin(\omega t+\alpha) \quad [13.27]$$

For a simply supported beam, we have an ideal admissible function which gives an exact solution to this problem.

$$\varphi_i = \sin(i\pi x / L) \quad [13.28]$$

The potential energy expression in equation [13.11] has two terms that depend on the flexural displacement w. These are associated with the flexural strain and the axial force effect which is a function of the axial force and the slope of the beam and are therefore identified with the subscripts elastic and axial force as follows:

$$V_{elastic} = \int_0^L \frac{EI}{2}\left(\frac{\partial^2 w}{\partial x^2}\right)^2 dx \quad [13.29]$$

$$V_{axial\_force} = -\int_0^L \frac{P}{2}\left(\frac{\partial w}{\partial x}\right)^2 dx \quad [13.30]$$

From Table 6.1, the corresponding stiffness matrix terms will be given as follows:

$$\text{Elastic stiffness coefficient } K_{i,j} = \int_0^L EI\varphi_i''\varphi_j'' dx \quad [13.31]$$

$$\text{Geometric stiffness coefficient } K'_{i,j} = -\int_0^L P\varphi_i'\varphi_j' dx \quad [13.32]$$

As the axial force is constant, this may be rearranged to give:

$$K'_{i,j} = -PK^* \quad [13.33a]$$

$$\text{where } K^* = \int_0^L \varphi'_i \varphi'_j dx \quad [13.33b]$$

The mass matrix is the same as for the beams without any axial force which we have considered and is given by:

$$M_{i,j} = \int_0^L \rho A \varphi_i \varphi_j dx \quad [13.34]$$

The eigenvalue equation can then be formulated as follows:

$$[K]\{G\}-P[K^*]\{G\}-\omega^2[M]\{G\}=\{0\} \quad [13.35]$$

If P is known (or prescribed), the two stiffness matrices could be added and the square of the frequency w is the eigenvalue.

$$[K_{overall}]\{G\}-\omega^2[M]\{G\}=\{0\} \quad [13.36]$$

$$\text{where } [K_{overall}]= [K]-P[K^*] \quad [13.37]$$

Thus the solution to the above equation obviously yields the natural frequencies and modes of an axially loaded beam subject to a prescribed static axial compressive force. Tensile force can be accommodated by changing the sign of P. Another interesting point top observe is that if we remove the mass matrix in equation [13.35], and consider P as an unknown, this then gives us the eigenvalue equation for finding the critical loads of the beam. Another way to look at the critical state is to recognize that if a calculated natural frequency from the solution of equation [13.35] turned out to be zero then the implication is that the equilibrium is in a critical state and the load P is a critical load.

We chose simply supported beams for a good reason because these functions are orthogonal giving diagonal matrices for the elastic stiffness, geometric stiffness and the mass.

$$\varphi'_i = \sin(i\pi x / L) = (i\pi / l)\cos(i\pi x / L) \quad [13.38]$$

$$\varphi''_i = \sin(i\pi x / L) = -(i\pi / l)^2 \sin(i\pi x / L) \quad [13.39]$$

It is easy to see that with the sinusoidal shape for the admissible function defined in equation [13.28],

$$\text{As } \int_0^L \varphi_i \varphi_j dx = \int_0^L \varphi_i' \varphi_j' dx = \int_0^L \varphi_i'' \varphi_j'' dx = 0 \text{ for } i \neq j$$

All matrices in equation [13.35] are diagonal.

Also since $\int_0^L \sin^2(i\pi x / L)dx = \int_0^L \cos^2(i\pi x / L)dx = L/2$, the diagonal terms are then given by:

$$K_{i,i} = (i\pi / L)^4 EI(L/2) \quad [13.40]$$

$$K^*_{i,i} = (i\pi / L)^2 (L/2) \quad [13.41]$$

$$K_{i,i} = \rho A(L/2) \quad [13.42]$$

Thus the ith row of the eigenvalue equation gives the uncoupled equation:

$$\left((i\pi / L)^{2\,4} EI(L/2) - P(i\pi / L)^2 (L/2) - \omega^2 \rho A(L/2)\right) G_i = 0$$

For non-trivial solution,

$$\left((i\pi / L)^4 EI - P(i\pi / L)^2 - \omega^2 \rho A\right) = 0 \quad [13.43]$$

This gives the following expression for ith natural frequency,

$$\omega = \sqrt{\frac{i^4 \pi^4 EI}{\rho AL^4} - \frac{i^2 \pi^2 P}{\rho AL^2}} \quad [13.44]$$

The expression for the ith critical load is obtained by setting the above expression to zero (corresponding to zero natural frequency) which for the first mode gives the well-known Euler load P_E.

$$\frac{i^4 \pi^4 EI}{\rho AL^4} - \frac{i^2 \pi^2 P}{\rho AL^2} = 0 \Rightarrow P_{crit,i} = \frac{i^2 \pi^2 EI}{L^2} = i^2 P_E \quad [13.45]$$

13.5. The point of no return: zero natural frequency

13.5.1. *Natural frequency*

All structures that are in a state of *stable, static equilibrium* have a definite configuration (shape). If such a structure that is at rest were subject to a force or disturbance, that configuration would change. If the structure is then released by removing the force, it would tend to return to its original equilibrium configuration. The following example illustrates this point.

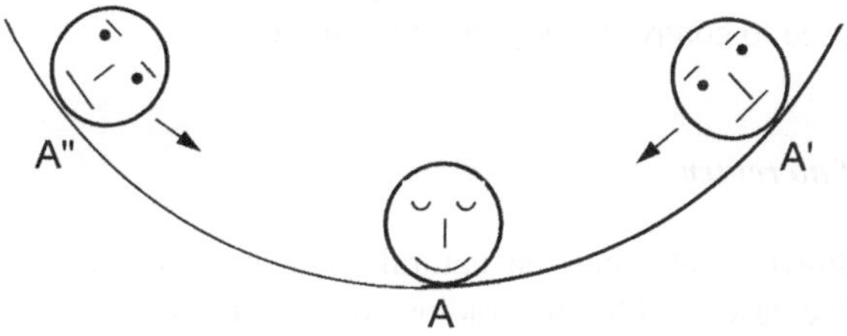

Figure 13.5. *Stable equilibrium*

A ball resting (let us say "comfortably") at the bottom (A) of a concave surface (see Figure 13.5) is in a state of stable, static equilibrium. If it is displaced to a point A′ and released, it would move toward A. It would approach A with a velocity, and therefore it would pass this point and move further on the other side till it comes to a momentary halt at point A″. If there are no frictional forces or air resistance, A′ and A″ would be at the same level, and the ball would again move in the opposite direction, passing A to reach point A′ again. The ball will always have an acceleration toward the equilibrium state A. In the absence of friction or air resistance, this vibratory motion can go on indefinitely, the ball moving between A′ and A″. The speed at which the ball passes A is proportional to the *frequency* of this oscillation. If there are no external dynamic (time dependent) forces (as in this case), the motion is said to be a free vibration. In this example, although a force is required initially to displace the ball, once released it vibrates freely. In this system, the motion of the ball takes place along one path only. This is an example of a *single degree of freedom* system. This free vibration can take place only at a particular value of frequency, which is its *natural frequency.* The *return period* of vibration is the time taken to complete a full cycle, and is equal to the inverse of the natural frequency in cycles per second.

13.5.2. *Why not forever?*

In reality, nothing goes on forever, except perhaps examinations. Any frictional force, which behaves like an opposition in parliament, causes energy loss during vibration, and the motion will eventually cease. Such forces that absorb energy are classified as *damping forces*, and the vibration in the presence of these forces is called a *damped free vibration.* As there will always be some energy loss, an undamped free vibration exists only in theory and all free vibrations in real life have an end. Nevertheless, in many practical situations the damping forces are not significantly large enough to alter the natural frequencies noticeably, and in such cases the frequencies may be calculated by using an undamped model. However, the amplitude of vibration is sensitive to damping, and for this reason artificial dampers are sometimes used to control the vibration of structures.

13.5.3. *Point of no return*

It can be shown easily that the natural frequency of the ball in the bowl in Figure 13.5 decreases with the radius of curvature of the bowl. As the surface becomes flatter, the frequency diminishes. When it is perfectly flat, the natural frequency will be zero. This means the return period is infinity. The ball would never return to its original equilibrium state (A) if displaced and released at any arbitrary point A′ (see Figure 13.6). It would stay in the displaced position, in a state of equilibrium. The equilibrium configuration is therefore indefinite. Such a state of equilibrium is referred to as a *critical* or *neutral state of equilibrium*.

Figure 13.6. *Critical equilibrium*

A third type of equilibrium called *unstable equilibrium* is illustrated in Figure 13.7. Any disturbance to this system will initiate a motion, away from the original equilibrium state. There will be no oscillation, and the system does not have a natural frequency. Mathematically, the frequency may take an imaginary value.

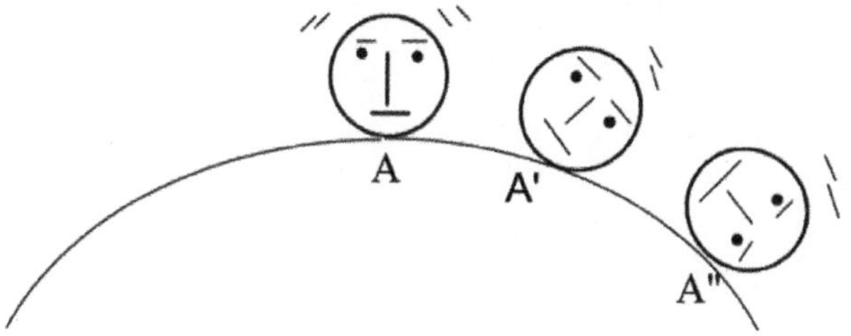

Figure 13.7. *Unstable equilibrium*

14

The RRM in Finite Elements Method

14.1. Discretization of structures

While the Rayleigh–Ritz method (RRM) is a versatile procedure for structural analysis, for structures with complicated geometry it is useful to divide the structure into a large number of small elements and derive the mathematical equations at certain nodal points along directions in which the structure has the ability to move. This is called the Finite Element Method (FEM). The displacements along these nodes are referred to as the degrees of freedom. The RRM can then be applied to each element in terms of the degrees of freedom which are similar to the undetermined coefficients associated with admissible functions in the general RRM. The admissible functions for the finite structural elements are formulated so that the set of functions possess all essential degrees of freedom at the nodes. The functions used to represent the displacements of elements are called shape functions or interpolation functions in FEM. They are commonly denoted by N.

14.2. Theoretical basis

To illustrate the theoretical basis of the method, let us consider a beam element. The deflection of the beam $f(x)$ is the field variable. There are two ways to derive the finite element matrix equations. One is based on the principle of minimum potential energy, which may be regarded as a Rayleigh–Ritz procedure. Consider the general case of a beam of length L, flexural rigidity EI under a distributed loading of intensity $q(x)$ and a point load F applied at $x = \beta L$ and a concentrated moment M_0 applied at $x = \gamma L$.

The potential energy of the system is given by the following expression which needs to be minimized:

$$V = \int_0^L (EI/2)(f''(x))^2 dx - \int_{x=0}^L q(x)f(x)dx - F.f(\beta L) - M_0.f'(\gamma L) \quad [14.1]$$

Here, both deflection f and load intensity q are assumed to be in the same direction. Since the force is doing work while moving in the direction of displacement potential energy is lost and hence the negative sign.

The other is the solution of the governing differential equation obtained using equation of equilibrium. This means solving the differential equation:

$$EI\frac{d^4 f}{dx^4} - q(x) = 0 \quad [14.2]$$

This equation may be solved using Galerkin's method. We will see later that this formulation would also result in integrals of products of second derivative just as in the energy method. This means the field variable must be a function that can be differentiated twice. In order to allow this to vary at least linearly, we need to have a linear function for the second derivative. Therefore, if we chose polynomials, we should have constant, linear, quadratic and cubic terms.

Whichever method we choose to use, we will start with the derivation of the same shape function which is common to both approaches.

14.3. Essential conditions at the boundaries and nodes

In the beam problem, the potential energy terms require second derivatives of the displacement. This means that any assumed function for the field variable must be differentiable twice throughout the system and that the first derivatives must be continuous at the nodes connecting any two elements. If this is not the case then the second derivative would be indefinite at the nodes. This means having an infinite curvature and hence infinite bending moment which is not achievable. The implication is that at each node the displacement f as well as the flexural rotation f' must be continuous. Such a function is called C_1 continuous. C_n continuous function is one whose nth derivatives are continuous at the nodes. So what does this mean for a beam? To maintain continuity at the nodes the translations and rotations of each element must be equated to those of the adjacent elements. The displacement within each element is expressed as functions of the nodal values of the translation and rotation.

14.4. Derivation of interpolation functions (shape functions)

Consider a typical beam element say rth element. Let the length of the beam be $L_{(r)}$ and the flexural rigidity be $EI_{(r)}$.

Let us chose a series

$$f_{(r)}(x) = \sum_{j=1,2}^{4} \hat{d}_{j(r)} N_j \quad [14.3]$$

where $\hat{d}_{j(r)}$ are the four end displacements (translation of the left end, rotation of the left end, translation of the right end and the rotation of the right end) of the rth beam element and N are the shape functions. This notation is necessary to derive the element properties but it should be remembered that these end displacements correspond to the nodal displacements of the overall structure and need to be identified by the global numbering when the final matrix is assembled. In the example shown in section 14.5, the rotation of the right end of the first beam element $\hat{d}_{4(1)}$ is the same as the rotation of the left end of the second element $\hat{d}_{2(2)}$ and is labeled d_1 in the overall structural notation, i.e. for the example under consideration, $\hat{d}_{4(1)} = \hat{d}_{2(2)}) = d_1$.

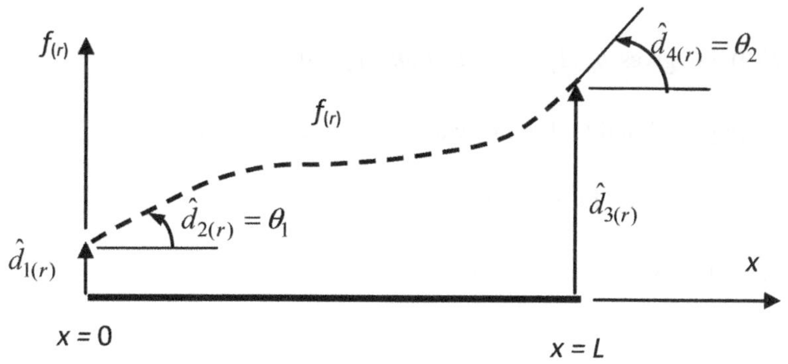

Figure 14.1. *Beam element*

The element displacement must then satisfy the following relationships:

$$f_r(0) = \hat{d}_{1(r)};\; f_r'(0) = \hat{d}_{2(r)};\; f_r(L_r) = \hat{d}_{3(r)};\; f_r'(L_r) = \hat{d}_{4(r)}$$

These would hold if the shape function had the following properties:

$$N_1(0)=1; N_2(0)=0; N_3(0)=0; N_4(0)=0;$$
$$N_1'(0)=0; N_2'(0)=1; N_3'(0)=0; N_4'(0)=0;$$
$$N_1(L_r)=0; N_2(L_r)=0; N_3(L_r)=1; N_4(L_r)=0;$$
$$N_1'(L_r)=0; N_2'(L_r)=0; N_3'(L_r)=0; N_4'(L_r)=1;$$

The task is then to design shape functions that satisfy all the above equations. Let us start with the first shape.

Let $N_1 = a_1 + b_1(x/L_r) + c_1(x/L_r)^2 + h_1(x/L_r)^3$

Then $N_1' = b_1(1/L_r) + c_1(2x/L_r^2) + h_1(3x^2/L_r^3)$

$N_1(0)=1$ gives $a_1 = 1$

$N_1'(0)=0$ gives $b_1 = 0$

$N_1(L_r)=0$ gives $a_1 + b_1 + c_1 + h_1 = 0$

Using the values found for the first two coefficients, $1 + c_1 + h_1 = 0$

$N_1'(0)=0$ gives $b_1 / L_k + 2c_1 / L_k + 3h_1 / L_k = 0$

Substituting a_1=1 and b_1=0 into the above gives $c_1 = -1.5h_1$.

But we already have $1 + c_1 + h_1 = 0$.

Therefore $1 - 1.5h_1 + h_1 = 0$ giving $h_1 = 2$.

Therefore $c_1 = -3$.

$$N_1 = 1 - 3(x/L_r)^2 + 2(x/L_r)^3.$$

Similarly other shape functions can be determined. The result is:

$$N_1 = 1 - 3(x/L_r)^2 + 2(x/L_r)^3$$

$$N_2 = L_r(x/L_r) - 2L_r(x/L_r)^2 + L_r(x/L_r)^3$$

$$N_3 = 3(x/L_r)^2 - 2(x/L_r)^3$$

$$N_4 = -L_r(x/L_r)^2 + L_r(x/L_r)^3$$

These may also be written as:

$$\{N\} = \begin{bmatrix} 1 & 0 & -3 & 2 \\ 0 & L_r & -2L_r & L_r \\ 0 & 0 & 3 & -2 \\ 0 & 0 & -L_r & L_r \end{bmatrix} \begin{Bmatrix} 1 \\ (x/L_r) \\ (x/L_r)^2 \\ (x/L_r)^3 \end{Bmatrix} \quad [14.4]$$

$$f_r(x) = \sum_{j=1,2}^{4} f_{j(r)} N_j \quad [14.5]$$

or in matrix notation $f_r(x) = \{N\}^{\mathrm{T}} \{\hat{d}_r\}$ [14.6]

EXERCISE.– verify that the above formulation gives the required nodal displacements.

14.5. Derivation of element matrix equations using the Rayleigh–Ritz method

$$V = \int_0^L (EI/2)(f''(x))^2 dx - \int_{x=0}^L q(x) f(x) dx - F.f(\beta L) - M_0.f'(\gamma L) \quad [14.7]$$

For equilibrium the potential energy must be a minimum. To minimize the above functional we need to differentiate it with some adjustable parameters. So far, we have defined the shape functions but the nodal displacements or degrees of freedom have not been determined. Therefore, we can minimize with respect to these and find the particular set of nodal displacements that would satisfy the equilibrium as closely as possible.

The equation is then:

$$\left\{\frac{\partial V}{\partial d_i}\right\} = \{0\} \quad [14.8]$$

where d_i are the nodal displacements. Once again it is worth remembering that for each element there are only four nodal displacements and the subscript i gives the nodal degree of freedom number for the overall structure but for each applicable element this would correspond to one of the four end displacements.

$$d_i = \hat{d}_{i(r)},\ i \in [1,2,3,4] \qquad [14.9]$$

Considering the contribution from each element to the minimization equation, we can differentiate the energy within an element with respect to its four nodal displacements and set up the elements of the matrix equation and finally assemble the terms to form the overall minimization equation. Consider the potential energy within the rth element

$$V_{(r)} = \int_0^{L_r} (EI_r/2)\left(f_r''(x)\right)^2 dx - \int_{x=0}^{L_r} q(x) f_r(x) dx - Ff_r(\beta L_r) - M_0 f_r'(\gamma L_r)$$

$$\begin{aligned} &= \int_0^{L_r} (EI_r/2)\left(d_{1(r)}N_1'' + d_{2(r)}N_2'' + d_{3(r)}N_3'' + d_{4(r)}N_4''\right)^2 dx \\ &- \int_0^{L_k} q(x)\left(d_{1(r)}N_1 + d_{2(r)}N_2 + d_{3(r)}N_3 + d_{4(r)}N_4\right) dx \\ &- F\left(d_{1(r)}N_1(bL_r) + d_{2(r)}N_2(bL_r) + d_{3(r)}N_3(bL_r) + d_{4(r)}N_4(bL_r)\right) \\ &- M_0\left(d_{1(r)}N_1'(cL_r) + d_{2(r)}N_2'(cL_r) + d_{3(r)}N_3'(cL_r) + d_{4(r)}N_4'(cL_r)\right) \end{aligned} \qquad [14.10]$$

Let us differentiate with respect to a typical nodal displacement $\delta_{k,i}$ which gives:

$$\frac{\partial V_{(r)}}{\partial d_{i(r)}} = \int_0^{L_r} \left(\frac{EI_r}{2}\right) 2N_i''\left(\sum_{j=1}^{4} N_j'' d_{j(r)}\right) dx - \int_0^{L_r} q(x) N_i dx - FN_i(\beta L_r) - M_0 N_i'(\gamma L_r) \qquad [14.11]$$

It should be remembered that the above equation contribute to the overall minimization equation but in itself it is not zero.

It may be rewritten as:

$$\frac{\partial V_{(r)}}{\partial d_{i(r)}} = \int_0^{L_r} EI_r N_i''\left(\sum_{j=1}^{4} N_j'' d_{j(r)}\right) dx - \int_0^{L_r} q(x) N_i dx - FN_i(\beta L_r) - M_0 N_i'(\gamma L_r) \qquad [14.12]$$

The order of summation and integration can be changed. Thus:

$$\frac{\partial V_{(r)}}{\partial d_{i(r)}} = \sum_{j=1}^{4} d_{j(r)} \int_0^{L_r} EI_r (N_i'' N_j'') dx - \int_0^{L_r} q(x) N_i dx - FN_i(\beta L_r) - M_0 N_i'(\gamma L_r) \qquad [14.13]$$

$$\text{Let } \int_0^{L_k} EI_k(N_i''N_j'')dx = K_{k,i,j} \qquad [14.14]$$

Then the above equations may be written as:

$$\frac{\partial V}{\partial d_{i(r)}} = \sum_{j=1}^{4} d_{j(r)}\hat{k}_{i,j(r)} - \hat{q}_{k,i} \qquad [14.15]$$

where $\hat{k}_{i,j(r)}$ is a stiffness coefficient associated with the *i*th and *j*th degrees of freedom of the *r*th element. It can be shown that it gives the action (force or moment) along the *i*th degree of freedom of the *r*th element required to cause a unit displacement (translation or rotation) along its *j*th degree of freedom. $Q_{i(r)}$is an equivalent nodal force due to the effect of the loading on the *i*th degree of freedom and is called the generalized force (this could be a force or a moment).

The stiffness coefficients give rise to a stiffness matrix which is a property of the structure while the generalized force depends on the imposed loading.

For each element there will be four such equations which may be written in matrix form as:

$$\begin{Bmatrix} \dfrac{\partial V}{\partial \hat{d}_{1(r)}} \\ \dfrac{\partial V}{\partial \hat{d}_{2(r)}} \\ \dfrac{\partial V}{\partial \hat{d}_{3(r)}} \\ \dfrac{\partial V}{\partial \hat{d}_{4(r)}} \end{Bmatrix} = \begin{bmatrix} \hat{k}_{1,1(r)} & \hat{k}_{1,2(r)} & \hat{k}_{1,3(r)} & \hat{k}_{1,4(r)} \\ \hat{k}_{2,1(r)} & \hat{k}_{2,2(r)} & \hat{k}_{2,3(r)} & \hat{k}_{2,4(r)} \\ \hat{k}_{3,1(r)} & \hat{k}_{3,2(r)} & \hat{k}_{3,3(r)} & \hat{k}_{3,4(r)} \\ \hat{k}_{4,1(r)} & \hat{k}_{4,2(r)} & \hat{k}_{4,3(r)} & \hat{k}_{4,4(r)} \end{bmatrix} \begin{Bmatrix} \hat{d}_{1(r)} \\ \hat{d}_{2(r)} \\ \hat{d}_{3(r)} \\ \hat{d}_{4(r)} \end{Bmatrix} - \begin{Bmatrix} \hat{q}_{1(r)} \\ \hat{q}_{2(r)} \\ \hat{q}_{3(r)} \\ \hat{q}_{4(r)} \end{Bmatrix} \qquad [14.16]$$

It may be shown that for the shape functions derived the element stiffness matrix of a uniform beam made of homogeneous material is:

$$\left[\hat{k}_{(r)}\right] = \begin{bmatrix} 12EI_r/L_r^3 & 6EI_r/L_r^2 & -12EI_r/L_r^3 & 6EI_r/L_r^2 \\ 6EI_r/L_r^2 & 4EI_r/L_r & -6EI_r/L_r^2 & 2EI_r/L_{rk} \\ -12EI_r/L_r^3 & -6EI_r/L_r^2 & 12EI_r/L_r^3 & -6EI_r/L_r^2 \\ 6EI_r/L_r^2 & 2EI_r/L_r & -6EI_r/L_r^2 & 4EI_r/L_r \end{bmatrix} \qquad [14.17]$$

One again note that the matrix is symmetrical.

The generalized forces may be found for the following cases separately:

a) Uniform distributed load $q(x) = q_0$.

b) Point load F applied at $x=\beta L$.

c) Concentrated moment M_0 at $x=\gamma L$.

These are known as fixed end forces and fixed end moments in structural analysis. It should be pointed out here that in Finite Element Analysis (FEA), as we can introduce nodes at points where concentrated forces and moments are applied, such loads are treated as external nodal loads and therefore cases (b) and (c) are not commonly used.

14.5.1. *Uniform distributed load*

$$\begin{aligned}\hat{q}_{1(r)} &= \int_0^{L_r} q_0 N_1 dx = q_0 \int_0^{L_r} (1-3(x/L_r)^2 + 2(x/L_r)^3)dx \\ &= q_0\left[(x - x^3/L_r^2 + x^4/2L_r^3)\right]_0^{L_k} = q_0 L_r/2\end{aligned} \quad [14.18]$$

$$\begin{aligned}\hat{q}_{2(r)} &= \int_0^{L_r} q_0 N_2 dx = q_0 \int_0^{L_r} (L_r(x/L_r) - 2L_r(x/L_r)^2 + L_r(x/L_r)^3)dx \\ &= q_0\left[(x^2/2 - 2x^3/3L_r + x^4/4L_k^2)L_r\right]_0^{L_r} = q_0 L_r^2/12\end{aligned} \quad [14.19]$$

$$\begin{aligned}\hat{q}_{3(r)} &= \int_0^{L_r} q_0 N_3 dx = q_0 \int_0^{L_r} (3(x/L_r)^2 - 2(x/L_r)^3)dx \\ &= q_0\left[(x^3/L_r^2 - x^4/2L_r^3)\right]_0^{L_r} = q_0 L_r/2\end{aligned} \quad [14.20]$$

$$\begin{aligned}\hat{q}_{4(r)} &= \int_0^{L_r} q_0 N_4 dx = q_0 \int_0^{L_r} (-L_r(x/L_r)^2 + L_r(x/L_r)^3)dx \\ &= q_0\left[(-x^3/3L_r + x^4/4L_r^2)L_r\right]_0^{L_r} = -q_0 L_r^2/12\end{aligned} \quad [14.21]$$

14.5.2. *Point load*

A point load may be treated as a delta function included in the distributed load function q. i.e. a large force acting over a very short length which amounts to evaluating the shape function at the point of application of load and multiplying by the load.

$$\hat{q}_{1(r)} = F.N_1(\beta L_r) = F(1-3(\beta)^2 + 2(\beta)^3) \quad [14.22]$$

$$\hat{q}_{2(r)} = F.N_2(\beta L_r) = FL_r(\beta - 2\beta^2 + \beta^3) \quad [14.23]$$

$$\hat{q}_{3(r)} = F.N_3(\beta L_r) = F(3\beta^2 - 2\beta^3) \quad [14.24]$$

$$\hat{q}_{4(r)} = F.N_4(\beta L_r) = FL_r(-\beta^2 + \beta^3) \quad [14.25]$$

14.5.3. *Concentrated moment*

$$\hat{q}_{1(r)} = M_0.N_1'(\gamma L_r) = 6M_0\gamma(-1+\gamma)/L_r \quad [14.26]$$

$$\hat{q}_{2(r)} = M_0.N_2'(\gamma L_r) = M_0(1-4\gamma+3\gamma^2) \quad [14.27]$$

$$\hat{q}_{3(r)} = M_0.N_3'(\gamma L_r) = 6M_0\gamma(1-\gamma)/L_r \quad [14.28]$$

$$\hat{q}_{4(r)} = M_0.N_4'(\gamma L_r) = M_0\gamma(-2+3\gamma) \quad [14.29]$$

14.5.4. *External loads at the nodes*

If a point load F is applied at a node, then $\beta = 0$ or 1.

If the load is applied at the left end, $\beta = 0$. This gives $\hat{q}_{1(r)1} = F$ but all other equivalent actions will be zero.

Similarly, if $\beta = 1$, $\hat{q}_{3(r)} = F$ but all other equivalent actions will be zero.

This is equivalent to applying an additional force at the node. It is important to ensure that this load is applied only once. The load may be regarded as being applied at the end of one beam at a node but it should not be reapplied at other beams at the node.

Similarly we can show that a concentrated moment may be applied either at a node or using the equivalent nodal load formula on *one* of the beams at a node. For example, if $\gamma=0$, a moment is applied at the left end of a beam element. The corresponding nodal actions are all zero except $\hat{q}_{2(r)}$ which is equal to the applied moment M_0. For $\gamma = 1$, there will be a moment M_0 at the right end giving zero terms for all nodal actions except $\hat{q}_{4(r)}$ which is equal to the applied moment M_0.

It is the total energy that must be minimized. The above formulations must therefore be assembled to equate the total resisting actions to the applied external actions along all degrees of freedom.

14.6. Assembly of element matrices

Consider, as an example, the beam in Figure 14.2. It is clamped at one end and simply supported at another point and is subject to distributed and point loads as shown. The beam has a flexural rigidity *EI* and length *L*.

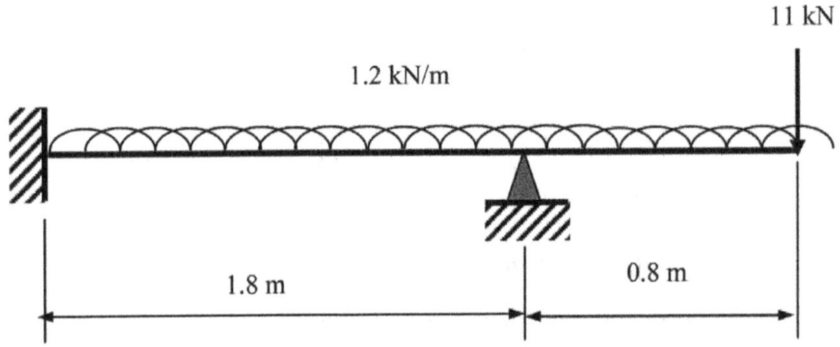

Figure 14.2. *Continuous beam*

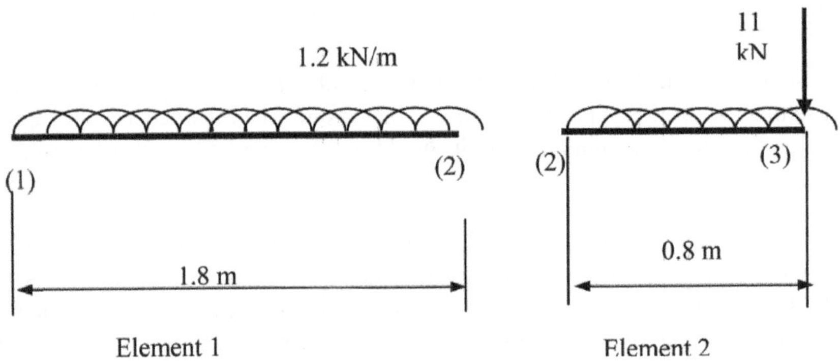

Figure 14.3. *Discretised continuous beam*

In assembling the matrices, it should be noted that each row of the final matrix equation gives an equilibrium statement to the effect that the sum of the forces along a particular degree of freedom is zero. As mentioned earlier, the degrees of freedom for a standard beam problem consist of nodal translations as well as rotations. The sum of all stiffness terms associated with a particular degree of freedom gives the total internal resistance and the corresponding sum of forces (or moments) gives the effect of the external loads (forces or moments). The summation is done over all the elements sharing a common degree of freedom. Let us illustrate this by solving the beam problem in Figure 14.2. It is necessary to have a node where there is a constraint. Therefore the beam is divided into two elements as shown in Figure 14.3. A constraint is a condition that makes the filed variable (or in some cases its derivatives if this gives rise to an essential condition as explained earlier) zero. At the left end where the beam is clamped there is a node. This is labelled (1). Since there is a simple support there should be a node at this point which is labelled (2). Node 3 is at the right end. With the three nodes there are 2 elements. Each element has two nodes one at the left end and the other at the right end. At each node there are two degrees of freedom: one translation and one flexural rotation (slope). So we will treat the structure as having potentially up to six degrees of freedom. However, the left end of the beam is clamped and the beam can neither translate nor rotate at this end. The common node cannot translate because of the simple support and has only one degree of freedom – a rotation. So we have three degrees of freedom. The degrees of freedom may be numbered globally as d_1, d_2, d_3 corresponding to the rotation of the common node, the translation of the right end and the rotation of the right end respectively.

FEM enables us to formulate the governing equations, in this case, equations of equilibrium, in terms of the degrees of freedom, the structural properties and the loading. The structural properties are obtained in terms of its stiffness defined by a matrix called the stiffness matrix and the effect of loading is expressed as a vector which depends on the loading and the degrees of freedom chosen.

The overall matrix equation is:

$$\begin{bmatrix} K_{1,1} & K_{1,2} & K_{1,3} \\ K_{2,1} & K_{2,2} & K_{2,3} \\ K_{3,1} & K_{2,2} & K_{3,3} \end{bmatrix} \begin{Bmatrix} d_1 \\ d_2 \\ d_3 \end{Bmatrix} = \begin{Bmatrix} Q_1 \\ Q_2 \\ Q_3 \end{Bmatrix} \qquad [14.30]$$

Once the stiffness matrix $[K]$ and the force vector $\{Q\}$ are determined the displacements $\{d\}$ along the degrees of freedom may be found by solving the above set of simultaneous equations. The stiffness of the structure is obtained by summing the stiffness contribution of each element. For example, $K_{2,2}$ is the moment required to produce a unit rotation of the beam at the common node which depends on the stiffness of both elements. The element stiffness coefficients can be formulated conveniently in terms of locally defined degrees of freedom for each element.

Here the nodal displacement d_1 is shared by both elements and the appropriate stiffness coefficients should be added. Thus the first row and first column involve both members.

$K_{1,1}$ is the moment due to a unit rotation. Using the element stiffness matrix, we now have the following:

$$K_{1,1} = K_{1,4,4} + K_{2,2,2} = \frac{4EI}{L_1} + \frac{4EI}{L_2}$$

$$K_{1,2} = K_{2,2,3} = -\frac{6EI}{L_2^2}$$

$$K_{1,3} = K_{2,2,4} = \frac{2EI}{L_2}$$

Note that there is no effect from the left end of the first beam as it is constrained fully.

$$K_{2,2} = K_{2,3,3} = \frac{12EI}{L_2^3}$$

$$K_{2,3} = K_{2,3,4} = -\frac{6EI}{L_2^2}$$

$$K_{3,3} = K_{2,4,4} = \frac{4EI}{L_2}$$

The generalized forces may also be found by assembling the member end forces.

The overall stiffness matrix is:

$$\begin{bmatrix} \left(\frac{4EI}{L_1} + \frac{4EI}{L_2}\right) & \left(-\frac{6EI}{L_2^2}\right) & \left(\frac{2EI}{L_2}\right) \\ \left(-\frac{6EI}{L_2^2}\right) & \left(\frac{12EI}{L_2^3}\right) & \left(-\frac{6EI}{L_2^2}\right) \\ \left(\frac{2EI}{L_2}\right) & \left(-\frac{6EI}{L_2^2}\right) & \left(\frac{4EI}{L_2}\right) \end{bmatrix} \begin{Bmatrix} d_1 \\ d_2 \\ d_3 \end{Bmatrix} = \begin{Bmatrix} Q_1 \\ Q_2 \\ Q_3 \end{Bmatrix} \qquad [14.31]$$

Substituting the element lengths gives:

$$\begin{bmatrix} (7.5EI) & (-9.375EI) & (2.5EI) \\ (-9.375EI) & (23.4375EI) & (-9.375EI) \\ (2.5EI) & (-9.375EI) & (5EI) \end{bmatrix} \begin{Bmatrix} d_1 \\ d_2 \\ d_3 \end{Bmatrix} = \begin{Bmatrix} Q_1 \\ Q_2 \\ Q_3 \end{Bmatrix} \quad [14.32]$$

It may be noted that the overall stiffness matrix is also symmetrical.

The nodal forces for the global structure can also be formulated in the same way.

$$Q_1 = \hat{q}_{4,(1)} + \hat{q}_{2,(2)}$$

$$Q_2 = \hat{q}_{3,(2)}$$

$$Q_3 = \hat{q}_{4,(2)}$$

Using equations [14.19–21],

$$Q_1 = -(1.2)\times 1.8^2 / 12 + (1.2)\times 0.8^2 / 12 = \text{-0.26 kNm}$$

$$Q_2 = (1.2)\times 0.8 / 2 = 0.48 \text{ kN}$$

$$Q_3 = -(1.2)\times 0.8^2 / 12 = 0.064 \text{ kNm.}$$

With these, equation [14.32] can be solved to find the degrees of freedom and these in turn could then be used to find forces and moments in the members along the element degrees of freedom.

We will now derive the system matrices to solve some simple eigenvalue problems.

14.7. Eigenvalue problems: geometric stiffness matrix for calculating critical loads

The potential energy of a beam subject to a static compressive load P can be shown to be:

$$V = \int_0^L (EI/2)\left(f''(x)\right)^2 dx - \int_{x=0}^L (P/2)(f'(x))^2 dx \quad [14.33]$$

The application of the RRM with the shape functions derived earlier will lead to the eigenvalue equation:

$$[K]\{d\}-P[K']\{d\}=\{0\}, \qquad [14.34]$$

where the coefficients of the geometric stiffness matrix $[K']$ for an element are given by:

$$[K']=\begin{bmatrix} 1.2/L & 1/10 & -1.2/L & 1/10 \\ 1/10 & 2L/15 & -1/10 & -L/30 \\ -1.2/L & -1/10 & 1.2/L & -1/10 \\ 1/10 & -L/30 & -1/10 & 2L/15 \end{bmatrix} \qquad [14.35]$$

The critical loads for determined by solving the eigenvalue matrix equation. The eigenvectors give the buckling modes.

14.8. Eigenvalue problems: vibration analysis

The energy functional that needs to be minimized in a dynamic analysis of slender beams is of the form:

$$V=\int_0^L (EI/2)(f''(x))^2 dx-\omega^2\int_{x=0}^L (m/2)(f(x))^2 dx \qquad [14.36]$$

Here, m is the mass per unit length. The application of the RRM with the shape functions derived earlier will lead to the eigenvalue equation:

$$[K]\{d\}-\omega^2[M]\{d\}=\{0\}, \qquad [14.37]$$

where the coefficients of the mass matrix $[M]$ for an element are given by:

$$\hat{m}_{i,j}=\int_0^L mN_iN_j dx \qquad [14.38]$$

14.9. Consistent mass matrix for a beam element

Equation [14.38] leads to an element mass matrix called the *consistent mass matrix.*

$$[\hat{m}]=\begin{bmatrix} 13mL/35 & 11mL^2/210 & 9mL/70 & -13mL^2/420 \\ 11mL^2/210 & mL^3/105 & 13mL^2/420 & -mL^3/140 \\ 9mL/70 & 13mL^2/420 & 13mL/35 & -11mL^2/210 \\ -13mL^2/420 & -mL^3/140 & -11mL^2/210 & mL^3/105 \end{bmatrix} \qquad [14.39]$$

14.10. Lumped mass matrix for a beam element

Another approach used in dynamic analysis is to assume the masses as concentrated at the nodes. This called the lumped mass model. This means each translational degree of freedom would be associated with a mass equal to half the mass of the beam. This results in a diagonal matrix which is easier to formulate but the solution is less accurate.

$$[\hat{m}] = \begin{bmatrix} mL/2 & 0 & 0 & 0 \\ 0 & 0 & 0 & 0 \\ 0 & 0 & mL/2 & 0 \\ 0 & 0 & 0 & 0 \end{bmatrix} \qquad [14.40]$$

Note that the rotary inertia of the beam is neglected in the slender beam theory. This is not accurate for higher modes even for the case of slender beams. To include the effect of the rotary inertia an additional energy terms needs to be included. This is:

$$\omega^2 \int_{x=0}^{L} (mr^2/2)(f'(x))^2\,dx$$

where r is the radius of gyration about the neutral axis. It should be mentioned here that for problems where the rotary inertia has a significant contribution (deep beams) the shear deformation effect also needs to be considered.

14.11. The Rayleigh–Ritz and the Galerkin methods

With problems for which a variational statement of the type we have been dealing with (based on conservation of energy for example) is not available, a different approach is required. For such problems we would seek to solve a governing differential equation. A common approach is a weighted residual method called the Galerkin method. It is worth commenting on how this method is applied and compare with the RRM for a problem we have already considered.

The governing equilibrium of a beam element is:

$$EI_r \frac{d^4 f_r}{dx^4} - q(x) = 0 \qquad [14.41]$$

The idea is to solve this approximately by a weighted average method. Using the shape functions as the weighting functions and integrating the weighted residual over the domain of the element length gives the following equation:

$$\int_0^{L_r} \left(EI_r \frac{d^4 f_r}{dx^4} - q(x) \right) N_i dx = 0 \qquad [14.42]$$

The second term is identical to the corresponding term in the Rayleigh–Ritz approach. The first term requires fourth derivative of the shape functions. The cubic functions are not differentiable to this extent. However, we can integrate this expression by parts and reduce the order of differentiation. This is called the weak formulation.

$$\int_0^{L_r} \left(EI_r \frac{d^4 f_r}{dx^4} \right) N_i dx = -\int_0^{L_r} \left(EI_r \frac{d^3 f_r}{dx^3} \right) N_i' dx + \left[EI_r \frac{d^3 f_r}{dx^3} N_i \right]_0^{L_r}$$

$$= + \int_0^{L_r} \left(EI_r \frac{d^2 f_k}{dx^2} \right) N_i'' dx - \left[EI_k \frac{d^2 f_r}{dx^2} N_i' \right]_0^{L_r} + \left[EI_r \frac{d^3 f_r}{dx^3} N_i \right]_0^{L_r} \qquad [14.43]$$

The first term leads to the same stiffness matrix as in the RRM. The second and third terms are boundary values (in a two dimensional (2D) case they would be boundary integrals) consisting of internal actions (bending moments and shear forces) multiplied by corresponding displacements. It may be shown that at all internal nodes the internal actions will cancel each other and at the boundaries of the overall structure either the displacement or the action would be zero. If the boundary conditions are not fully clamped or completely free, the boundary may be replaced by an element possessing equivalent stiffness in which case the above arguments would still hold. This shows that both Galerkin's method and the RRM give the same matrix equations.

The above formulation does not directly give the effect of concentrated forces and moments but they can be treated as special cases using delta functions, although as mentioned already, it is better to introduce nodes at the points and consider the applied loads as external nodal loads.

Bibliography

[ASK 10] ASKES H., CARAMÉS-SADDLER M., RODRÍGUEZ-FERRAN A., "*Bipenalty method for time domain computational dynamics*", *Proceedings of the Royal Society A: Mathematical, Physical and Engineering Sciences*, vol. 466, pp. 1389–1408, 2010.

[AVI 98] AVILÉS R., AJURIA G., BILBAO A., *et al.*, "Lagrange multipliers and the primal-dual method in the non-linear static equilibrium of multibody systems", *Communications in Numerical methods in Engineering*, vol. 14, pp. 463–472, 1998.

[BAT 96] BATHE K.J., *Finite element procedures*, Englewood Cliffs, N.J., Prentice Hall, 1996.

[BER 94] BERT C.W., MALIK M., "Frequency equations and modes of free vibrations of rectangular plates with various edge conditions", *Proceedings of the Institution of Mechanical Engineers, Part C: Journal of Mechanical Engineering Science*, vol. 208, pp. 307–319, 1994.

[BHA 85] BHAT R.B., "Natural frequencies of rectangular plates using characteristic orthogonal polynomials in Rayleigh-Ritz method", *Journal of Sound and Vibration*, vol. 102, pp. 493–499, 1985.

[BRO 97] BROWN R.E., STONE M.A., "On the use of polynomial series with the Rayleigh-Ritz method", *Composite Structures*, vol. 39, pp. 191–196, 1997.

[BUD 46] BUDIANSKY B., HU P.C., The Lagrangian multiplier method of finding upper and lower limits to critical stresses of clamped plates, NACA report 848, 1946.

[CHE 92] CHENG L., NICOLAS J., "Free vibration analysis of a cylindrical shell-circular plate system with general coupling and various boundary conditions", *Journal of Sound and Vibration*, vol. 155, pp. 231–247, 1992.

[CRO 97] CROSSLAND J.A., DICKINSON S.M., "The free vibration of thin rectangular planform shallow shells with slits", *Journal of Sound and Vibration*, vol. 199, pp. 513–521, 1997.

[COU 43] COURANT R., "Variational methods for the solution of problems of equilibrium and vibrations", *Bulletin of the American Mathematics Society*, vol. 49, pp. 1–23, 1943.

[COU 53] COURANT R., HILBERT D., *Methods of Mathematical Physics*, Interscience Publishers Inc., New York, vol. I, p. 175, 1953.

[DOW 84] DOWELL E.H., "On asymptotic approximations to beam model shapes", *Journal of Applied Mechanics, Transactions ASME*, vol. 51, 439, 1984.

[FRA 05] FRASER C, "Joseph Louis Lagrange,Théorie des fonctions analytiques", *Landmark Writings in Western Mathematics 1640–1940*, Ed. Ivor Grattan-Guinness. Elsevier, pp. 208–224, 2005.

[GAV 00] GAVETE L., BENITO J.J., FALCON S., *et al.*, "Implementation of essential boundary conditions in a meshless method", *Communications in Numerical methods in Engineering*, vol. 16, pp. 409–421, 2000.

[GOR 89] GORMAN D.G., "A comprehensive study of the free vibration of rectangular plates resting on symmetrically-distributed uniform elastic edge supports", *Journal of Applied Mechanics*, vol. 56, pp. 839–899, 1989.

[GOU 66] GOULD S.H., *Variational Methods for Eigenvalue Problems: An Introduction to the Weinstein Method of Intermediate Problems*, London, Oxford University Press, vol. 10, 1966.

[HET 13] HETHERINGTON J., RODRÍGUEZ-FERRAN A., ASKES H., "The bipenalty method for arbitrary multipoint constraints", *International Journal for Numerical Methods in Engineering*, vol. 93, pp. 465–482, 2013.

[ILA 86] ILANKO S., DICKINSON S.M., "Linearization of the strain energy expressions for a vibrating slightly curved beam subjected to an axial load", *Journal of Sound and Vibration*, vol. 108, pp. 357–359, 1986.

[ILA 90] ILANKO S., "The vibration behaviour of initially imperfect simply supported beams subject to axial loading", *Journal of Sound and Vibration*, vol. 142, pp. 355–359, 1990.

[ILA 99] ILANKO S., DICKINSON S.M., "Asymptotic modeling of rigid boundaries and connections in the Rayleigh-Ritz method", *Journal of Sound and Vibration*, vol. 219, pp. 370–378, 1999.

[ILA 00a] ILANKO S., asymptot.exe, Interactive program available on line as a zipped archive at Ilanko.org/asymptot.zip, 2000.

[ILA 00b] ILANKO S., netonian.exe, Interactive program available on line as a zipped archive at Ilanko.org/newtonian.zip, 2000.

[ILA 02a] ILANKO S., "Existence of natural frequencies of systems with imaginary Restraints and their convergence in asymptotic modeling", *Journal of Sound and Vibration*, vol. 255 no. 5, pp. 883–898, 2002.

[ILA 02b] ILANKO S., "The use of negative penalty functions in constrained variational problems", *Communications in Numerical Methods in Engineering*, vol. 18, no. 9 pp. 659–668, 2002.

[ILA 05] ILANKO S., "Introducing the use of positive and negative inertial functions in asymptotic modelling", *Proceedings of the Royal Society*, vol. 461, pp. 2545–2562, 2005.

[ILA 08] ILANKO S., WILLIAMS F.W, WITTRICK -W., "Algorithm proof of bracketing and convergence theorems for eigenvalues of constrained structures with positive and negative penalty parameters", *International Journal for Numerical Methods in Engineering*, vol. 7, no. 75, pp. 83–102, 2008.

[ILA 09] ILANKO S., "Comments on the historical bases of the Rayleigh and Ritz methods", *Journal of Sound and Vibration*, vol. 319, pp. 731–733, 2009.

[ILA 12] ILANKO S., MONTERRUBIO L.E., "Bipenalty method from a frequency domain perspective", *International Journal for Numerical Methods in Engineering*, vol. 90, pp. 1278–1291, 2012.

[KAR 00] KARNOVSKY I.A, LEBED O.I., *Formulas for Structural Dynamics: Tables, Graphs & Solutions*, McGraw-Hill Professional Book Group, 2000.

[KIM 90] KIM C.S., YOUNG P.G., DICKINSON S.M., "On the flexural vibration of rectangular plates approached by using simple polynomials in the Rayleigh-Ritz Method", *Journal of Sound and Vibration*, vol. 142, pp. 143–394, 1990.

[LAG 01] LAGRANGE J.L., "Leçons sur le calcul des fonctions", *Séances des Ecoles Normales*, vol. 10, pp. 1–534, 1801.

[LEE 94] LEE H.P., NG T.Y., "*Natural* frequencies and modes for the flexural vibration of a cracked beam", *Applied Acoustics*, vol. 42, pp. 151–163, 1994.

[LEI 73] LEISSA A.W., "The free vibration of rectangular plates", *Journal of Sound and Vibration*, vol. 31, pp. 257–293, 1973.

[LEI 84] LEISSA A.W., NARITA Y., "Vibrations of completely free shallow shells of rectangular planform". *Journal of Sound and Vibration*, vol. 96, pp. 207–218, 1984.

[LEI 05] LEISSA A.W., "The historical bases of the Rayleigh and Ritz methods", *Journal of Sound and Vibration*, vol. 287, pp. 961–978, 2005.

[LI 00] LI W.L., "Free vibrations of beams with general boundary conditions", *Journal of Sound and Vibration*, vol. 237, pp. 709–725, 2000.

[LI 02] LI W.L., DANIELS M., "A Fourier series method for the vibration of elastically restrained plates arbitrarily loaded with springs and masses", *Journal of Sound and Vibration*, vol. 252, pp. 768–781, 2002.

[LI 04] LI W.L., "Vibration analysis of rectangular plates with general elastic boundary supports", *Journal of Sound and Vibration*, vol. 273, pp. 619–635, 2004.

[LIE 93] LIEW K.M., HUNG K.C., LIM M.K., "A continuum three-dimensional vibration analysis of thick rectangular plates", *International Journal of Solids and Structures*, vol. 30, 1993.

[MCC 73] MCCALLION H., *Vibration of Mechanical Systems*, Longmans, London, 1973.

[MEI 86] MEIROVITCH L., *Elements of Vibration Analysis*, Boston, McGraw-Hill, 1986.

[MOC 12] MOCHIDA Y., ILANKO S., DUKE M., *et al.*, "Free vibration analysis of doubly curved shallow shells using the Superposition-Galerkin method", *Journal of Sound and Vibration*, vol. 331, pp. 1413–1425, 2012.

[MON 09] MONTERRUBIO L.E., "Free vibration of shallow shells using the Rayleigh–Ritz method and penalty parameters", *Proceedings of the Institution of Mechanical Engineers, Part C: Journal of Mechanical Engineering Science*, vol. 23, pp. 2263–2272, 2009.

[MON 12a] MONTERRUBIO L.E., "The use of eigenpenalty parameters in structural stability analysis", *Proceedings of the Institution of Mechanical Engineers, Part C: Journal of Mechanical Engineering Science*, vol. 226, pp. 861–870, 2012.

[MON 12b] MONTERRUBIO L.E., "Frequency and buckling parameters of box-type structures", *Computers & Structures*, vol. 104–105, pp. 44–49, 2012.

[MUK 08] MUKHOPADHYAY M., "Structural dynamics: vibration and systems", *Ane Books India for CRC Press*, New Delhi, India, 2008.

[NAG 08] NAGINO H., MIKAMI T., MIZUSAWA T., "Three-dimensional free vibration analysis of isotropic rectangular plates using the B-spline Ritz method", *Journal of Sound and Vibration*, vol. 317, 329–353, 2008.

[OOS 95] OOSTERHOUT G.M., VAN DER HOOGT P.J.M., SPIERING R.M.E.J., "Accurate calculation methods for natural frequencies of plates with special attention to the higher modes", *Journal of Sound and Vibration*, vol. 183, pp. 33–47, 1995.

[QAT 12] QATU M.S., ASADI E., "Vibration of doubly curved shallow shells with arbitrary boundary conditions", *Applied Acoustics*, vol. 73, pp. 21–27, 2012.

[RAY 73] LORD RAYLEIGH J.W.S., "On the nodal lines of a square plate", *Philosophical Magazine* Series 4, vol. 46, pp. 166–171, 1873.

[RAY 99] LORD RAYLEIGH J.W.S., "On the calculation of the frequency of vibration in its gravest mode, with an example from hydrodynamics", *Philosophical Magazine*, pp. 566–572, 1899.

[RAY 11] LORD RAYLEIGH J.W.S., "On the calculation of Chladni's figures for a square plate", *Philosophical Magazine,* Sixth Series, vol. 22, pp. 225–229, 1911.

[RAY 45a] LORD RAYLEIGH J.W.S., *The Theory of Sound*, The Macmillan Company, vol. 1, 1877. (Reprinted 1945 by Dover Publications, New York).

[RAY 45b] LORD RAYLEIGH J.W.S., *The Theory of Sound*, The Macmillan Company, vol. 2, 1878. (Reprinted 1945 by Dover Publications, New York).

[RIT 08] RITZ W., "U¨ ber eine neue Methode zur Lo¨ sung gewisser Variationsprobleme der mathematischen Physik", *Journal fu¨r Reine und Angewandte Mathematik*, vol. 135, pp. 1–61, 1908.

[RIT 09] RITZ W., "Theorie der Transversalschwingungen einer quadratische Platte mit freien Ra¨ ndern", *Annalen der Physik*, vol. 28, pp. 737–786, 1909.

[SCH 85] SCHÄFER B., "Free vibrations of a gravity-loaded clamped-free bea", *Ingenieur-Archiv.* vol. 55, pp. 66–80, 1985.

[SOU 69] SOUTHWELL R.V., *An Introduction to The Theory of Elasticity for Engineers and Physicists*, Dover Publications Inc., New York,1969.

[SZI 04] SZILARD R., *Theories and Applications of Plate Analysis: Classical, Numerical, and Engineering Methods*, John Wiley, Hoboken, NJ, USA, 2004.

[TEM 33] TEMPLE B.F., BICKLEY W.G., *Rayleigh's Principle and its Applications to Engineering*, Oxford University Press, London, 1933.

[THO 88] THOMSON W.T., *Theory of Vibration with Applications,* 3rd ed., Unwin Publications, London, 1988.

[TIM 55] TIMOSHENKO S., YOUNG D.H., *Vibration Problems in Engineering,* 3rd ed., Van Nostrand Publications, New York, 1955.

[VEN 01] VENTSEL E., KRAUTHAMMER T., *Thin Plates and Shells – Theory, Analysis, and Applications*, Marcel Dekker, Inc., USA, 2001.

[WAR 94] WARKENTIN A., SEMERCIGIL S.E., "A passive joint controller for the flexible robotic arm", *Journal of Sound and Vibration*, vol. 174, no. 1, pp. 123–134, 1994.

[WU 07] WU J.H., LIU A.Q., CHEN H.L., "Exact solutions for free-vibration analysis of rectangular plates using Bessel functions", *Journal of Applied Mechanics*, vol. 74, pp. 1247–1251, 2007.

[YOK 88] YOKOYOMA T., "Free vibration characteristics of rotating Timoshenko beams", *International Journal Mechanical. Science,* vol. 30, no. 10, 1988, pp. 743–755.

[YOU 50] YOUNG D., "Vibration of rectangular plates by the Ritz method," Journal of Applied Mechanics, vol. 17, pp. 448–453, 1950.

[YOU 95] YOUNG P.G., DICKINSON S.M., "*Vibration of a class of shallow shells bounded by edges described by polynomials, part I: theoretical approach and validation*", *Journal of Sound and Vibration*, vol. 181, 1995, pp. 203–214.

[YUA 92a] YUAN J., DICKINSON S.M., "On the use of artificial springs in the study of the free vibrations of systems comprised of straight and curved beams", *Journal of Sound and Vibration*, vol. 153, pp. 203–216, 1992.

[YUA 92b] YUAN J., DICKINSON S.M., "The flexural vibration of rectangular plate systems approached by using artificial springs in the Rayleigh-Ritz method", *Journal of Sound and Vibration*, vol. 159, pp. 39–55, 1992.

[YUF 09] YUFENG X., LIU B., "New exact solutions for free vibrations of rectangular thin plates by simplectic dual method", *Acta Mechanica Sinica*, vol. 25, pp. 265–270, 2009.

[ZHO 96] ZHOU D., "Natural frequencies of rectangular plates using a set of static beam functions in Rayleigh-Ritz method", *Journal of Sound and Vibration*, vol. 189, pp. 81–87, 1996.

[ZIE 74] ZIENKIEWICZ O.C., *Constrained Variational Principles and Penalty Function Methods in the Finite Element Analysis*, Lecture Notes in Mathematics, vol. 363, Springer, Berlin, pp. 207–214, 1974.

[ZIE 77] ZIENKIEWICZ O.C., *The Finite Element Method*, Third Edition, Mc-Graw-Hill, London, pp. 284–287, 1977.

Appendix

A.1. Rayleigh-Ritz with stiffness and mass penalty for a cantilever beam with polynomial admissible functions

```
clear all
n=9;
kp=-10^5;
mp=0;
for i=1:n
    for j=1:n
        K(i,j)=kp+(i*j*(i+1)*(j+1)/(i+j-1));
        M(i,j)=1/(i+j+3)+mp;
    end
end
[mode,omega]=eig(K,M);
for i=1:n
    w(i)=(omega(i,i))^0.25;
end

w(2)
w(3)
w(4)
```

A.2. Lagrangian Multiplier Method for a cantilever beam with polynomial admissible functions

```
clear all
n=9;
kp=0;
mp=0;
```

```
for i=1:n
    for j=1:n
        K(i,j)=kp+(i*j*(i+1)*(j+1)/(i+j-1));
        M(i,j)=1/(i+j+3)+mp;
    end
    K(n+1,i)=1;
    K(i,n+1)=1;
    M(n+1,i)=0;
    M(i,n+1)=0;
end
K(n+1,n+1)=0;
M(n+1,n+1)=0;
[mode,omega]=eig(K,M);
for i=1:n
    w(i)=(omega(i,i))^0.25;
end


w(n)
w(n-1)
w(n-2)
w(n-3)
```

A.3. RRM for calculating the critical loads and modes of cantilever or propped cantilever beams with polynomial admissible functions

```
clear all
n=9;
kp=10^6;
mp=0;
nm=100;
for i=1:n
    for j=1:n
        K(i,j)=kp+(i*j*(i+1)*(j+1)/(i+j-1));
        M(i,j)=(i+1)*(j+1)/(i+j+1)+mp;
    end
end
[mode,omega]=eig(K,M);
for i=1:n
    w(i)=(omega(i,i));
end
w(1)
w(2)
w(3)
```

```
w(4)
for i=1:nm+1
    x(i)=(i-1)/nm;
    xx=x(i);
    mode1(i)=0;
    mode2(i)=0;
    mode3(i)=0;
    mode4(i)=0;

    for j=1:n
        mode1(i)=mode1(i)+mode(j,1)*xx^(j+1);
        mode2(i)=mode2(i)+mode(j,2)*xx^(j+1);
        mode3(i)=mode3(i)+mode(j,3)*xx^(j+1);
        mode4(i)=mode4(i)+mode(j,4)*xx^(j+1);
    end
end

plot(x,mode1)
hold on
plot (x,mode2)
hold on
plot(x,mode3)
hold on
plot (x,mode4)
```

A.4. RRM for calculating the natural frequencies and modes of vibration of beams using admissible functions presented in Chapter 8

```
format long
clear
clc
close all
%%%%%%%%%%%%%%%%%%%%%%%%%%%%%%%%%%%%%%%%%%%%%%%%%%%%%%%%%
%FREQUENCY PARAMETERS AND MODES OF EULER_BERNOULLI BEAMS
%   Luis Monterrubio                        2007.11.13
%%%%%%%%%%%%%%%%%%%%%%%%%%%%%%%%%%%%%%%%%%%%%%%%%%%%%%%%%
% Number of terms
nmax=13;
n=nmax;
TL=1;
% Defining the type of Penalty Parameter
cM=0; % Inertial Penalty Parameter
cK=1; % Stiffness type Penalty Parameter
% Penalty Parameter Value
```

```
PenaltyParameter=1*10^6;
%%%%%%%%%%%%%%%%%%%%%%%%%%%%%%%%%%%%%%%%%%%%%%%%%%%%%%%
%               DEFINING A BEAM FROM x=0 to x=L
%%%%%%%%%%%%%%%%%%%%%%%%%%%%%%%%%%%%%%%%%%%%%%%%%%%%%%%
% Defining Boundary Conditions
k1=PenaltyParameter*1;  % Translation at x=0
k2=PenaltyParameter*1;  % Rotation at x=0
k3=PenaltyParameter*0;  % Translation at x=1
k4=PenaltyParameter*0;  % Rotation at x=1
% Getting the Mass and Stiffness matrix of the beam
[K,M,KC] = EigKMBeams(nmax,k1,k2,k3,k4);
%%%%%%%%%%%%%%%%%%%%%%%%%%%%%%%%%%%%%%%%%%%%%%%%%%%%%%%
%                   SOLVING THE EIGENVALUE PROBLEM
%%%%%%%%%%%%%%%%%%%%%%%%%%%%%%%%%%%%%%%%%%%%%%%%%%%%%%%
[X,Eigenvalues]= eig(K+KC*cK,M+KC*cM);
Eigenvalues=diag(Eigenvalues);
Non_Dimesional_Freq_Par_rad_per_sec=(Eigenvalues).^0.25
%%%%%%%%%%%%%%%%%%%%%%%%%%%%%%%%%%%%%%%%%%%%%%%%%%%%%%%
%  BUILDING THE EIGENVECTORS                          %
%%%%%%%%%%%%%%%%%%%%%%%%%%%%%%%%%%%%%%%%%%%%%%%%%%%%%%%
% Building a vector containing the location of the
% points evaluated along the axis "X".
nn=201;
xplot=0:(TL/(nn-1)):TL;
% Building a 3D array PHI, that contains the modal
% matrix defined as X(k,i) multiplied by the deflection
% shape function, which increases its series
% along the k axis of the array and is evaluated
% in the i axis of the array n nn number of points
% along the "X" axis of the beam,.
PHI(n,nn+1,n)=0;
for i=1:n
    for j=1:nn+1
        for k=1:n
            if k<=3
                if j==0
                    PHI(i,j,k)=1;
                else
                    PHI(i,j,k)=...
                    X(k,i)*((((j-1)/(nn-1)))^(k-1));
                end
            elseif k>=4
                if j==0
                    PHI(i,j,k)=1;
                else
```

```
                    PHI(i,j,k)=...
                X(k,i)*(cos((k-3)*((j-1)/(nn-1))*pi));
                end
            end
        end
    end
end
% Building the eigenvectors by carrying out the
% sumation of the 3D array
% PHI on its k axis as defined before, that contains
% the columns of the modal matrix (eigenvectors).
% The result is the Matrix DISP containing the
% sumation of the columns containing the eigenvectors
% multiplied by the corresponding element of the series
% according to the row. Then, the DISP(i,j) matrix
% contains the evaluated eigenvectors in its
% rows i evaluated from L/nn to L.
DISP=zeros(n,nn);
for i=1:n
    for j=1:nn
        for k=1:n
            DISP(i,j)=DISP(i,j)+PHI(i,j,k);
        end
    end
end
%%%%%%%%%%%%%%%%%%%%%%%%%%%%%%%%%%%%%%%%%%%%%%%%%%%%%%%%%%
%  PLOTTING THE EIGENVECTORS                             %
%%%%%%%%%%%%%%%%%%%%%%%%%%%%%%%%%%%%%%%%%%%%%%%%%%%%%%%%%%
f = figure('Position',[50 50 900 600]);
plot(xplot,...
    [DISP(1,:)],xplot,[DISP(2,:)],xplot, ...
    [DISP(3,:)],xplot,[DISP(4,:)],xplot, ...
    [DISP(5,:)],xplot,[DISP(6,:)],xplot, ...
    [DISP(7,:)],xplot,[DISP(8,:)])
title(['Modes of Vibration'],'FontSize',24)
hx1=...
xlabel('Normalized Length of the Beam','FontSize',18);
hy1=ylabel('Deflection of the Beam','FontSize',18);
```

%Function EigKMBeams

```
function [K,M,KC] = EigKMBeams(nmax,k1,k2,k3,k4)
%%%%%%%%%%%%%%%%%%%%%%%%%%%%%%%%%%%%%%%%%%%%%%%%%%%%%%%%%%
% Stiffness Matrix of a beam
```

```
%%%%%%%%%%%%%%%%%%%%%%%%%%%%%%%%%%%%%%%%%%%%%%%%%%%%
K(3,3)=4;
for i=4:nmax
    K(i,i)=0.5*pi^4*(i-3)^4;
end
%%%%%%%%%%%%%%%%%%%%%%%%%%%%%%%%%%%%%%%%%%%%%%%%%%%%
% Adding springs or masses
%%%%%%%%%%%%%%%%%%%%%%%%%%%%%%%%%%%%%%%%%%%%%%%%%%%%
% TRANSLATIONAL SPRING AT X=0
Kk1(1,1)=1;
for i=4:nmax
    Kk1(1,i)=1;
    Kk1(i,1)=1;
end
for i=4:nmax
    for j=4:nmax
        Kk1(i,j)=1;
    end
end
% ROTATIONAL SPRING AT X=0
Kk2=zeros(nmax);
Kk2(2,2)=1;
% TRANSLATIONAL SPRING AT X=1
Kk3=ones(3);
for i=4:nmax
    Kk3(1,i)=cos((i-3)*pi);
    Kk3(i,1)=cos((i-3)*pi);
    Kk3(2,i)=cos((i-3)*pi);
    Kk3(i,2)=cos((i-3)*pi);
    Kk3(3,i)=cos((i-3)*pi);
    Kk3(i,3)=cos((i-3)*pi);
end
for i=4:nmax
    for j=4:nmax
        Kk3(i,j)=cos((i-3)*pi)*cos((j-3)*pi);
    end
end
% ROTATIONAL SPRING AT X=1
Kk4(2,2)=1;
Kk4(2,3)=2;
Kk4(3,2)=2;
Kk4(3,3)=4;
Kk4(nmax,nmax)=0;
%PENALTY MATRIX
KC=(k1*Kk1+k2*Kk2+k3*Kk3+k4*Kk4);
```

```
%%%%%%%%%%%%%%%%%%%%%%%%%%%%%%%%%%%%%%%%%%%%%%%%%%%%%%%%
%                         MASS MATRIX OF A BEAM
%%%%%%%%%%%%%%%%%%%%%%%%%%%%%%%%%%%%%%%%%%%%%%%%%%%%%%%%
M(1,1)=1;
M(1,2)=1/2;
M(2,1)=1/2;
M(1,3)=1/3;
M(2,2)=1/3;
M(3,1)=1/3;
M(2,3)=1/4;
M(3,2)=1/4;
M(3,3)=1/5;
for i=4:nmax
    M(2,i)=(cos((i-3)*pi)-1)/((i-3)^2*pi^2);
    M(i,2)=(cos((i-3)*pi)-1)/((i-3)^2*pi^2);
end
for i=4:nmax
    M(3,i)=2*cos((i-3)*pi)/((i-3)^2*pi^2);
    M(i,3)=2*cos((i-3)*pi)/((i-3)^2*pi^2);
end
for i=4:nmax
    M(i,i)=1/2;
end
```

A.5. RRM for calculating the natural frequencies and modes of vibration of plates using admissible functions presented in Chapter 8

```
%%%%%%%%%%%%%%%%%%%%%%%%%%%%%%%%%%%%%%%%%%%%%%%%%%%%%%%%
%        RECTANGULAR PLATES                            %
%   Luis Monterrubio        Date:2008.01.08           %
%   Comments: The Rayleigh-Ritz Method was            %
%   used together with the set of                     %
%           admissible functions:                     %
%     phi 1 = 1                                       %
%     phi 2 = x                                       %
%     phi 3 = x^2                                     %
%     phi 4 = cos(i-3)*pi*x/L                         %
%%%%%%%%%%%%%%%%%%%%%%%%%%%%%%%%%%%%%%%%%%%%%%%%%%%%%%%%
clear
clc
format long
close all
% Defining the number of terms used in both
% directions (X and Y)
```

```
nmax=10;
poisson=0.3;
% Defining the value of the penalty parameter
PenaltyParameter=1e9*1;
MPP=0;
KPP=0;
%%%%%%%%%%%%%%%%%%%%%%%%%%%%%%%%%%%%%%%%%%%%%%%%%%%%%%%%
%                                PLATE
%%%%%%%%%%%%%%%%%%%%%%%%%%%%%%%%%%%%%%%%%%%%%%%%%%%%%%%%
%%%%%%%%%%%%%%%%%%%%%%%%%%%%%%%%%%%%%%%%%%%%%%%%%%%%%%%%
% Defining material properties of the plate
%%%%%%%%%%%%%%%%%%%%%%%%%%%%%%%%%%%%%%%%%%%%%%%%%%%%%%%%
% Defining the  modulus of Young's
E = 1;%
% Defining the Poisson ratio
poisson=0.3;
% Defining the dimensions of the plate
a=1; % [m] dimension along the X axis
b=1; % [m] dimension along the Y axis
h=1;%0.1; % [m] thickness of the plate
alpha=a/b;
% Getting the flexural rigidity of the plate
D=E*h^3/(12*(1-poisson^2));
% Defining the boundary conditions of the Plate
EDGE1=2; % 0 Free  1 Simply supported 2 Clamped
EDGE2=2;
EDGE3=2;
EDGE4=2;
%%%%%%%%%%%%%%%%%%%%%%%%%%%%%%%%%%%%%%%%%%%%%%%%%%%%%%%%%
PLATE MATRICES
%%%%%%%%%%%%%%%%%%%%%%%%%%%%%%%%%%%%%%%%%%%%%%%%%%%%%%%%
% Defining the value of the penalty parameters
% on the sides
[Kt1, Kr1, Kt2, Kr2, Kt3, Kr3, Kt4, Kr4] = ...
PlatesBC(EDGE1,EDGE2, EDGE3, EDGE4,PenaltyParameter);
% Defining the Mass and Stiffness Matrices of the Plate
[K, M, Penalty_Matrix] = PlateMatrices(nmax,...
    alpha, Kt1, Kr1, Kt2, Kr2, ...
Kt3, Kr3, Kt4, Kr4, poisson);
% NON-DIMENSIONAL SOLUTION
[X,Eigenvalues]=...
eig(K+Penalty_Matrix*KPP,M+Penalty_Matrix*MPP);
Non_Dim_NatFreq=diag(Eigenvalues).^0.5
%%%%%%%%%%%%%%%%%%%%%%%%%%%%%%%%%%%%%%%%%%%%%%%%%%%%%%%%
%  BUILDING THE EIGENVECTORS
%%%%%%%%%%%%%%%%%%%%%%%%%%%%%%%%%%%%%%%%%%%%%%%%%%%%%%%%%%
```

```
for MODE_NUMBER=1:9;
step_x=1/49; step_y=1/49;
x_points=0:step_x:1; y_points=0:step_y:1;
steps_X=size(x_points,2); steps_Y=size(y_points,2);
EIGENVECTOR=real(X(:,MODE_NUMBER))';
for i=1:steps_X
    x=x_points(i);
    for j=1:steps_Y
         y=y_points(j);
        SET_X(1)=1;
        SET_X(2)=x;
        SET_X(3)=x^2;
        SET_Y(1)=1;
        SET_Y(2)=y;
        SET_Y(3)=y^2;
        for k=1:nmax-3
            SET_X(k+3)=cos(k*pi*x);
            SET_Y(k+3)=cos(k*pi*y);
        end
        counter=0;
        for n=1:nmax
            for m=1:nmax
                counter=counter+1;
                EVALUATED_FUNCTIONS(counter)=...
                    SET_X(m)*SET_Y(n);
            end
        end
        EVALUATED_MODE(i,j)=...
            -dot(EVALUATED_FUNCTIONS,EIGENVECTOR);
    end
end
pause
view(60, 30);
surf(x_points,y_points,EVALUATED_MODE)
end
```

Function PlatesBC

```
function [Kt1, Kr1, Kt2, Kr2, Kt3, Kr3, Kt4, Kr4] = ...
    PlatesBC(EDGE1,EDGE2, EDGE3, EDGE4,PenaltyParameter)
%%%%%%%%%%%%%%%%%%%%%%%%%%%%%%%%%%%%%%%%%%%%%%%%%%%%%
%DEFINING THE STIFFNESS COEFFICIENTS OF THE SPRINGS
%%%%%%%%%%%%%%%%%%%%%%%%%%%%%%%%%%%%%%%%%%%%%%%%%%%%%
```

```
% Defining the stiffness of the translational springs
% EDGE1
if EDGE1 == 0
    Kt1=0;
    Kr1=0;
end
if EDGE1 == 1
    Kt1=PenaltyParameter;
    Kr1=0;
end
if EDGE1 == 2
    Kt1=PenaltyParameter;
    Kr1=PenaltyParameter;
end
% EDGE2
if EDGE2 == 0
    Kt2=0;
    Kr2=0;
end
if EDGE2 == 1
    Kt2=PenaltyParameter;
    Kr2=0;
end
if EDGE2 == 2
    Kt2=PenaltyParameter;
    Kr2=PenaltyParameter;
end
% EDGE3
if EDGE3 == 0
    Kt3=0;
    Kr3=0;
end
if EDGE3 == 1
    Kt3=PenaltyParameter;
    Kr3=0;
end
if EDGE3 == 2
    Kt3=PenaltyParameter;
    Kr3=PenaltyParameter;
end
% EDGE4
if EDGE4 == 0
    Kt4=0;
    Kr4=0;
end
```

```
if EDGE4 == 1
    Kt4=PenaltyParameter;
    Kr4=0;
end
if EDGE4 == 2
    Kt4=PenaltyParameter;
    Kr4=PenaltyParameter;
end
```

Function PlateMatrices

```
function [Krs, Mrs, Penalty_Matrix] = ...
    PlateMatrices(nmax,alpha, Kt1, Kr1, Kt2, Kr2, ...
    Kt3, Kr3, Kt4, Kr4, poisson)
%%%%%%%%%%%%%%%%%%%%%%%%%%%%%%%%%%%%%%%%%%%%%%%%%%%%%%%%%
%                    MASS MATRIX
%%%%%%%%%%%%%%%%%%%%%%%%%%%%%%%%%%%%%%%%%%%%%%%%%%%%%%%%%
% The 4-dimensional mass matrix Mminj is obtained
% combining every term of Emi00 with everyterm of Fnj00
% as follows:
%   M(m,i,n,j) = Emi00(m,i) * Fnj00(n,j)
% The 2-dimensional mass matrix is obtained changing the
% subscripts minj for uv using the following
% relationships:
% u=i+(j-1)*nmax
% v=m+(n-1)*nmax
%%%%%%%%%%%%%%%%%%%%%%%%%%%%%%%%%%%%%%%%%%%%%%%%%%%%%%%%%
% Defining Emi00
%%%%%%%%%%%%%%%%%%%%%%%%%%%%%%%%%%%%%%%%%%%%%%%%%%%%%%%%%
Emi00(1,1)=1;
Emi00(1,2)=1/2;
Emi00(2,1)=1/2;
Emi00(1,3)=1/3;
Emi00(3,1)=1/3;
Emi00(2,2)=1/3;
Emi00(2,3)=1/4;
Emi00(3,2)=1/4;
Emi00(3,3)=1/5;
for i=4:nmax
    Emi00(2,i)=-(1-cos((i-3)*pi))/((i-3)^2*pi^2);
    Emi00(i,2)=-(1-cos((i-3)*pi))/((i-3)^2*pi^2);
    Emi00(3,i)=(2*cos((i-3)*pi))/((i-3)^2*pi^2);
    Emi00(i,3)=(2*cos((i-3)*pi))/((i-3)^2*pi^2);
end
```

```
for m=4:nmax
    Emi00(m,m)=1/2;
end
%%%%%%%%%%%%%%%%%%%%%%%%%%%%%%%%%%%%%%%%%%%%%%%%%%%%%%%%%
% Defining Fnj00
%%%%%%%%%%%%%%%%%%%%%%%%%%%%%%%%%%%%%%%%%%%%%%%%%%%%%%%%%
Fnj00=Emi00;
%%%%%%%%%%%%%%%%%%%%%%%%%%%%%%%%%%%%%%%%%%%%%%%%%%%%%%%%%
% Obtaining the 4-Dimensional Mass matrix Mminj
%%%%%%%%%%%%%%%%%%%%%%%%%%%%%%%%%%%%%%%%%%%%%%%%%%%%%%%%%
for m=1:nmax
    for i=1:nmax
        u=i+(m-1)*nmax;
        for n=1:nmax
            for j=1:nmax
                v=j+(n-1)*nmax;
                Mrs(u,v)=Emi00(m,n)*Fnj00(i,j);
            end
        end
    end
end
%%%%%%%%%%%%%%%%%%%%%%%%%%%%%%%%%%%%%%%%%%%%%%%%%%%%%%%%%
%                     STIFFNESS MATRIX                  %
%%%%%%%%%%%%%%%%%%%%%%%%%%%%%%%%%%%%%%%%%%%%%%%%%%%%%%%%%
% The 4-dimensional matrix is obtained combining every
% term of the following matrices as follows:
%   Kminj(m,i,n,j)= Emi22(m,i)*Fn00(i,j) +
%   alpha^4*Emi00(m,i)*Fnj22(n,j) +
%   poisson*alpha^2*Emi02(m,i)*Fnj20(n,j) +
%   poisson*alpha^2*Emi20(m,i)*Fnj02(n,j) +
%   2*(1-poisson)*alpha^2*Emi11(m,i)*Fnj11(n,j)
% The 2-dimensional stiffness matrix is obtained
% changing the subscriptsminj for uv using the following
% relationships:
% u=i+(j-1)*nmax
% v=m+(n-1)*nmax
%%%%%%%%%%%%%%%%%%%%%%%%%%%%%%%%%%%%%%%%%%%%%%%%%%%%%%%%%
% Defining Emi22(m,i) and Fnj22(n,j)
%%%%%%%%%%%%%%%%%%%%%%%%%%%%%%%%%%%%%%%%%%%%%%%%%%%%%%%%%
Emi22(3,3)=4;
for i=4:nmax
    Emi22(i,i)=0.5*(i-3)^4*pi^4;
end
Fnj22=Emi22;
```

```
%%%%%%%%%%%%%%%%%%%%%%%%%%%%%%%%%%%%%%%%%%%%%%%%%%%%%%%%%
% Defining Emi02(m,i) and Fnj02(n,j)
%%%%%%%%%%%%%%%%%%%%%%%%%%%%%%%%%%%%%%%%%%%%%%%%%%%%%%%%%
Emi02(1,3)=2;
Emi02(2,3)=1;
Emi02(3,3)=2/3;
for i=4:nmax
    Emi02(2,i)= 1-cos((i-3)*pi);
    Emi02(3,i)=-2*cos((i-3)*pi);
end
for i=4:nmax
    Emi02(i,i)=-0.5*(i-3)^2*pi^2;
end
Fnj02=Emi02;
%%%%%%%%%%%%%%%%%%%%%%%%%%%%%%%%%%%%%%%%%%%%%%%%%%%%%%%%%
% Defining Emi20(m,i) and Fnj20(n,j)
%%%%%%%%%%%%%%%%%%%%%%%%%%%%%%%%%%%%%%%%%%%%%%%%%%%%%%%%%
Emi20=Emi02';
Fnj20=Emi20;
%%%%%%%%%%%%%%%%%%%%%%%%%%%%%%%%%%%%%%%%%%%%%%%%%%%%%%%%%
% Defining Emi11(m,i) and Fmi11(n,j)
%%%%%%%%%%%%%%%%%%%%%%%%%%%%%%%%%%%%%%%%%%%%%%%%%%%%%%%%%
Emi11(2,2)=1;
Emi11(2,3)=1;
Emi11(3,2)=1;
Emi11(3,3)=4/3;
for i=4:nmax
    Emi11(2,i)=cos((i-3)*pi)-1;
    Emi11(i,2)=cos((i-3)*pi)-1;
    Emi11(3,i)=2*cos((i-3)*pi);
    Emi11(i,3)=2*cos((i-3)*pi);
end
for i=4:nmax
    Emi11(i,i)=0.5*(i-3)^2*pi^2;
end
Fnj11=Emi11;
% The 2-dimensional matrix K(u,v)
for m=1:nmax
   for i=1:nmax
      u=i+(m-1)*nmax;
      for n=1:nmax
         for j=1:nmax
            v=j+(n-1)*nmax;
            Krs(u,v)=Emi22(m,n)*Fnj00(i,j)+...
            alpha^4*Emi00(m,n)* Fnj22(i,j)+...
            poisson*alpha^2*Emi02(m,n)*Fnj20(i,j)+...
```

```
            poisson*alpha^2*Emi20(m,n)*Fnj02(i,j)+...
           2*(1-poisson)*alpha^2*Emi11(m,n)* Fnj11(i,j);
         end
      end
   end
end
%%%%%%%%%%%%%%%%%%%%%%%%%%%%%%%%%%%%%%%%%%%%%%%%%%%%%%%%
%              ADDING PENALTY PARAMETERS
%%%%%%%%%%%%%%%%%%%%%%%%%%%%%%%%%%%%%%%%%%%%%%%%%%%%%%%%
%%%%%%%%%%%%%%%%%%%%%%%%%%%%%%%%%%%%%%%%%%%%%%%%%%%%%%%%
%        TRANSLATIONAL PENALTY PARAMETERS
%%%%%%%%%%%%%%%%%%%%%%%%%%%%%%%%%%%%%%%%%%%%%%%%%%%%%%%%
%%%%%%%%%%%%%%%%%%%%%%%%%%%%%%%%%%%%%%%%%%%%%%%%%%%%%%%%
%                 EDGE1 & EDGE2 (X=0 & Y=0)
%%%%%%%%%%%%%%%%%%%%%%%%%%%%%%%%%%%%%%%%%%%%%%%%%%%%%%%%
% X(x)X(x) Terms (Evaluated part)
Ktransx0XmXi(1,1)=1;
for i=4:nmax
    Ktransx0XmXi(1,i)=1;
    Ktransx0XmXi(i,1)=1;
end
for i=4:nmax
    for j=4:nmax
        Ktransx0XmXi(i,j)=1;
    end
end
% Y(y)Y(y) Terms (Integral part)
Ktransx0YnYj=Emi00;
for m=1:nmax
   for i=1:nmax
      u=i+(m-1)*nmax;
      for n=1:nmax
         for j=1:nmax
             v=j+(n-1)*nmax;
             Ktransx0(u,v)=...
             Kt1*Ktransx0XmXi(m,n)*Ktransx0YnYj(i,j)+...
             Kt2*Ktransx0YnYj(m,n)*Ktransx0XmXi(i,j);
         end
      end
   end
end
%%%%%%%%%%%%%%%%%%%%%%%%%%%%%%%%%%%%%%%%%%%%%%%%%%%%%%%%
%                EDGE3 & EDGE4 (X=1 & Y=1)
%%%%%%%%%%%%%%%%%%%%%%%%%%%%%%%%%%%%%%%%%%%%%%%%%%%%%%%%
% XX Terms
```

```
KtransxlXmXi=ones(3);
for i=4:nmax
    KtransxlXmXi(1,i)=cos((i-3)*pi);
    KtransxlXmXi(i,1)=cos((i-3)*pi);
    KtransxlXmXi(2,i)=cos((i-3)*pi);
    KtransxlXmXi(i,2)=cos((i-3)*pi);
    KtransxlXmXi(3,i)=cos((i-3)*pi);
    KtransxlXmXi(i,3)=cos((i-3)*pi);
end
for i=4:nmax
    for j=4:nmax
        KtransxlXmXi(i,j)=cos((i-3)*pi)*cos((j-3)*pi);
    end
end
% YY Terms
Ktransx1YnYj=Emi00;
for m=1:nmax
   for i=1:nmax
      u=i+(m-1)*nmax;
      for n=1:nmax
         for j=1:nmax
             v=j+(n-1)*nmax;
             Ktransx1(u,v)= ...
             Kt3*Ktransx1XmXi(m,n)*Ktransx1YnYj(i,j)+...
             Kt4*Ktransx1YnYj(m,n)*Ktransx1XmXi(i,j);
         end
      end
   end
end
%%%%%%%%%%%%%%%%%%%%%%%%%%%%%%%%%%%%%%%%%%%%%%%%%%%%%%%%%%%
%             ROTATIONAL PENALTY PARAMETERS
%%%%%%%%%%%%%%%%%%%%%%%%%%%%%%%%%%%%%%%%%%%%%%%%%%%%%%%%%%%
%%%%%%%%%%%%%%%%%%%%%%%%%%%%%%%%%%%%%%%%%%%%%%%%%%%%%%%%%%%
%              EDGE1 & EDGE2 (X=0 & Y=0)
%%%%%%%%%%%%%%%%%%%%%%%%%%%%%%%%%%%%%%%%%%%%%%%%%%%%%%%%%%%
% X(x)X(x) Terms (Evaluated part)
Krotx0XmXi=zeros(nmax);
Krotx0XmXi(2,2)=1;
% Y(y)Y(y) Terms (Integral part)
Krotx0YnYj=Emi00;
for m=1:nmax
   for i=1:nmax
      u=i+(m-1)*nmax;
      for n=1:nmax
         for j=1:nmax
```

```
                v=j+(n-1)*nmax;
                Krotx0(u,v)=...
                Kr1*Krotx0XmXi(m,n)*Krotx0YnYj(i,j)+...
                Kr2*Krotx0YnYj(m,n)*Krotx0XmXi(i,j);
            end
        end
    end
end
%%%%%%%%%%%%%%%%%%%%%%%%%%%%%%%%%%%%%%%%%%%%%%%%%%%%%%%%%%%%
%        EDGE3 & EDGE4 (X=1 & Y=1)
%%%%%%%%%%%%%%%%%%%%%%%%%%%%%%%%%%%%%%%%%%%%%%%%%%%%%%%%%%%%
% X(x)X(x) Terms (Evaluated part)
Krotx1XmXi=zeros(nmax);
Krotx1XmXi(2,2)=1;
Krotx1XmXi(2,3)=2;
Krotx1XmXi(3,2)=2;
Krotx1XmXi(3,3)=4;
% Y(y)Y(y) Terms (Integral part)
Krotx1YnYj=Emi00;

for m=1:nmax
    for i=1:nmax
        u=i+(m-1)*nmax;
        for n=1:nmax
            for j=1:nmax
                v=j+(n-1)*nmax;
                Krotx1(u,v)=...
                Kr3*Krotx1XmXi(m,n)*Krotx1YnYj(i,j)+...
                Kr4*Krotx1YnYj(m,n)*Krotx1XmXi(i,j);
            end
        end
    end
end
Penalty_Matrix=Krotx0+Krotx1+Ktransx0+Ktransx1;
```

A.6. RRM for calculating the natural frequencies and modes of vibration of shells using admissible functions presented in Chapter 8

```
%%%%%%%%%%%%%%%%%%%%%%%%%%%%%%%%%%%%%%%%%%%%%%%%%%%%%%%%%%%%
% MATLAB Code to solve Shells using penalty parameters
% Luis Monterrubio                      Date: 2008.02.11
%                                       Updated: 2008.05.27
%%%%%%%%%%%%%%%%%%%%%%%%%%%%%%%%%%%%%%%%%%%%%%%%%%%%%%%%%%%
clear all
```

```
close all
clc
format long g
% Number of functions in each direction
nmax=15;
% Penalty Parameter Value
PenaltyParameter=1*1e9;
% Geometrical properties of the shell
a=1;
b=1;
h=b/100;
Rx=10;
Ry=10;
% Material properties
poisson=0.3;
% Boundary Conditions
Kt1=1;
Kt2=1;
Kt3=1;
Kt4=1;
Kr1=1;
Kr2=1;
Kr3=1;
Kr4=1;
%%%%%%%%%%%%%%%%%%%%%%%%%%%%%%%%%%%%%%%%%%%%%%%%%%%%%%%%
%MATRICES THAT ARE INDEPENDENT FROM GEOMETRIC PROP
%%%%%%%%%%%%%%%%%%%%%%%%%%%%%%%%%%%%%%%%%%%%%%%%%%%%%%%%
%%%%%%%%%%%%%%%%%%%%%%%%%%%%%%%%%%%%%%%%%%%%%%%%%%%%%%%%
% DEFINING THE MATRICES X?? AND Y?? TO OBTAIN ALL
% SUBMATRICES TO DEFINE THE MASS AND STIFFNESS MATRICES
%%%%%%%%%%%%%%%%%%%%%%%%%%%%%%%%%%%%%%%%%%%%%%%%%%%%%%%%
%%%%%%%%%%%%%%%%%%%%%%%%%%%%%%%%%%%%%%%%%%%%%%%%%%%%%%%%
% Defining X00
%%%%%%%%%%%%%%%%%%%%%%%%%%%%%%%%%%%%%%%%%%%%%%%%%%%%%%%%
X00(1,1)=1;
X00(1,2)=1/2;
X00(2,1)=1/2;
X00(1,3)=1/3;
X00(3,1)=1/3;
X00(2,2)=1/3;
X00(2,3)=1/4;
X00(3,2)=1/4;
X00(3,3)=1/5;
for i=4:nmax
    X00(2,i)=-(1-cos((i-3)*pi))/((i-3)^2*pi^2);
```

```
    X00(i,2)=-(1-cos((i-3)*pi))/((i-3)^2*pi^2);
    X00(3,i)=(2*cos((i-3)*pi))/((i-3)^2*pi^2);
    X00(i,3)=(2*cos((i-3)*pi))/((i-3)^2*pi^2);
end
for m=4:nmax
    X00(m,m)=1/2;
end
Y00=X00;
%%%%%%%%%%%%%%%%%%%%%%%%%%%%%%%%%%%%%%%%%%%%%%%%%%%%%%%%%%
% Defining X01
%%%%%%%%%%%%%%%%%%%%%%%%%%%%%%%%%%%%%%%%%%%%%%%%%%%%%%%%%%
X01(1,2)=1;
X01(1,3)=1;
X01(2,2)=1/2;
X01(2,3)=2/3;
X01(3,2)=1/3;
X01(3,3)=1/2;
for j=4:nmax
    X01(1,j)=cos((j-3)*pi)-1;
    X01(2,j)=cos((j-3)*pi);
    X01(3,j)=cos((j-3)*pi) - ...
    (2*(cos((j-3)*pi)-1))/((j-3)^2*pi^2);

    X01(j,3)=(2*(cos((j-3)*pi)-1))/((j-3)^2*pi^2);
end
for i=4:nmax
    for j=4:nmax
        if i == j
            X01(i,i) = 0;
        else
            X01(i,j) = +((j-3)/2) * ...
            ((cos((j-i)*pi)-1)/(j-i)+ ...
            (cos((j+i-6)*pi)-1)/(j+i-6) );
        end
    end
end
Y01=X01;
%%%%%%%%%%%%%%%%%%%%%%%%%%%%%%%%%%%%%%%%%%%%%%%%%%%%%%%%%%
% Defining X10
%%%%%%%%%%%%%%%%%%%%%%%%%%%%%%%%%%%%%%%%%%%%%%%%%%%%%%%%%%
X10=X01';
Y10=X10;
%%%%%%%%%%%%%%%%%%%%%%%%%%%%%%%%%%%%%%%%%%%%%%%%%%%%%%%%%%
% Defining X11
%%%%%%%%%%%%%%%%%%%%%%%%%%%%%%%%%%%%%%%%%%%%%%%%%%%%%%%%%%
```

```
X11(2,2)=1;
X11(2,3)=1;
X11(3,2)=1;
X11(3,3)=4/3;
for i=4:nmax
    X11(2,i)=cos((i-3)*pi)-1;
    X11(i,2)=cos((i-3)*pi)-1;
    X11(3,i)=2*cos((i-3)*pi);
    X11(i,3)=2*cos((i-3)*pi);
end
for i=4:nmax
    X11(i,i)=0.5*(i-3)^2*pi^2;
end
Y11=X11;
%%%%%%%%%%%%%%%%%%%%%%%%%%%%%%%%%%%%%%%%%%%%%%%%%%%%%%%
% Defining X02
%%%%%%%%%%%%%%%%%%%%%%%%%%%%%%%%%%%%%%%%%%%%%%%%%%%%%%%
X02(1,3)=2;
X02(2,3)=1;
X02(3,3)=2/3;
for i=4:nmax
    X02(2,i)= 1-cos((i-3)*pi);
    X02(3,i)=-2*cos((i-3)*pi);
end
for i=4:nmax
    X02(i,i)=-0.5*(i-3)^2*pi^2;
end
Y02=X02;
%%%%%%%%%%%%%%%%%%%%%%%%%%%%%%%%%%%%%%%%%%%%%%%%%%%%%%%
% Defining Emi20(m,i) and Fnj20(n,j)
%%%%%%%%%%%%%%%%%%%%%%%%%%%%%%%%%%%%%%%%%%%%%%%%%%%%%%%
X20=X02';
Y20=X20;
%%%%%%%%%%%%%%%%%%%%%%%%%%%%%%%%%%%%%%%%%%%%%%%%%%%%%%%
% Defining Emi22(m,i) and Fnj22(n,j)
%%%%%%%%%%%%%%%%%%%%%%%%%%%%%%%%%%%%%%%%%%%%%%%%%%%%%%%
X22(3,3)=4;
for i=4:nmax
    X22(i,i)=0.5*(i-3)^4*pi^4;
end
Y22=X22;
%%%%%%%%%%%%%%%%%%%%%%%%%%%%%%%%%%%%%%%%%%%%%%%%%%%%%%%
%                  MASS MATRIX
%%%%%%%%%%%%%%%%%%%%%%%%%%%%%%%%%%%%%%%%%%%%%%%%%%%%%%%
% The 4-dimensional mass matrix Mminj is obtained
```

```
% combining every term
% of X00 with everyterm of Fnj00 as follows:
%   M(m,i,n,j) = X00(m,i) * Fnj00(n,j)
% The 2-dimensional mass matrix is obtained
% changing the subscripts
% minj for uv using the following relationships:
% u=i+(j-1)*nmax
% v=m+(n-1)*nmax
%%%%%%%%%%%%%%%%%%%%%%%%%%%%%%%%%%%%%%%%%%%%%%%%%%%%%%%%
% Obtaining the 4-Dimensional Mass matrix Mminj
%%%%%%%%%%%%%%%%%%%%%%%%%%%%%%%%%%%%%%%%%%%%%%%%%%%%%%%%
for m=1:nmax
    for i=1:nmax
        u=i+(m-1)*nmax;
        for n=1:nmax
            for j=1:nmax
                v=j+(n-1)*nmax;
                MUU(u,v)=X00(m,n)*Y00(i,j);
            end
        end
    end
end
ZEROS= zeros(nmax^2);
M=[MUU ZEROS ZEROS; ZEROS MUU ZEROS; ZEROS ZEROS MUU];
%%%%%%%%%%%%%%%%%%%%%%%%%%%%%%%%%%%%%%%%%%%%%%%%%%%%%%%%
%        ADDING PENALTY PARAMETERS
%%%%%%%%%%%%%%%%%%%%%%%%%%%%%%%%%%%%%%%%%%%%%%%%%%%%%%%%
%%%%%%%%%%%%%%%%%%%%%%%%%%%%%%%%%%%%%%%%%%%%%%%%%%%%%%%%
%       TRANSLATIONAL PENALTY PARAMETERS
%%%%%%%%%%%%%%%%%%%%%%%%%%%%%%%%%%%%%%%%%%%%%%%%%%%%%%%%
%%%%%%%%%%%%%%%%%%%%%%%%%%%%%%%%%%%%%%%%%%%%%%%%%%%%%%%%
%         EDGE1 & EDGE2 (X=0 & Y=0)
%%%%%%%%%%%%%%%%%%%%%%%%%%%%%%%%%%%%%%%%%%%%%%%%%%%%%%%%
% X(x)X(x) Terms (Evaluated part)
Ktransx0XmXi(1,1)=1;
for i=4:nmax
    Ktransx0XmXi(1,i)=1;
    Ktransx0XmXi(i,1)=1;
end
for i=4:nmax
    for j=4:nmax
        Ktransx0XmXi(i,j)=1;
    end
end
% Y(y)Y(y) Terms (Integral part)
```

```
KtransxOYnYj=X00;
for m=1:nmax
    for i=1:nmax
        u=i+(m-1)*nmax;
        for n=1:nmax
            for j=1:nmax
                v=j+(n-1)*nmax;
                Ktransx0(u,v)=...
                Kt1*Ktransx0XmXi(m,n)* ...
                Ktransx0YnYj(i,j)+...
                Kt2*Ktransx0YnYj(m,n)* ...
                Ktransx0XmXi(i,j);
            end
        end
    end
end
%%%%%%%%%%%%%%%%%%%%%%%%%%%%%%%%%%%%%%%%%%%%%%%%%%%%%%%
%                        EDGE3 & EDGE4  (X=1 & Y=1)
%%%%%%%%%%%%%%%%%%%%%%%%%%%%%%%%%%%%%%%%%%%%%%%%%%%%%%%
% XX Terms
Ktransx1XmXi=ones(3);
for i=4:nmax
    Ktransx1XmXi(1,i)=cos((i-3)*pi);
    Ktransx1XmXi(i,1)=cos((i-3)*pi);
    Ktransx1XmXi(2,i)=cos((i-3)*pi);
    Ktransx1XmXi(i,2)=cos((i-3)*pi);
    Ktransx1XmXi(3,i)=cos((i-3)*pi);
    Ktransx1XmXi(i,3)=cos((i-3)*pi);
end
for i=4:nmax
    for j=4:nmax
        Ktransx1XmXi(i,j)=cos((i-3)*pi)*cos((j-3)*pi);
    end
end
% YY Terms
Ktransx1YnYj=X00;
for m=1:nmax
    for i=1:nmax
        u=i+(m-1)*nmax;
        for n=1:nmax
            for j=1:nmax
                v=j+(n-1)*nmax;
                Ktransx1(u,v)=...
                Kt3*Ktransx1XmXi(m,n)* ...
                Ktransx1YnYj(i,j)+...
```

```
                    Kt4*Ktransx1YnYj(m,n)* ...
                    Ktransx1XmXi(i,j);
                end
            end
        end
    end
%%%%%%%%%%%%%%%%%%%%%%%%%%%%%%%%%%%%%%%%%%%%%%%%%%%%%%%%
%             ROTATIONAL PENALTY PARAMETERS
%%%%%%%%%%%%%%%%%%%%%%%%%%%%%%%%%%%%%%%%%%%%%%%%%%%%%%%%
%%%%%%%%%%%%%%%%%%%%%%%%%%%%%%%%%%%%%%%%%%%%%%%%%%%%%%%%
%                         EDGE1 & EDGE2 (X=0 & Y=0)
%%%%%%%%%%%%%%%%%%%%%%%%%%%%%%%%%%%%%%%%%%%%%%%%%%%%%%%%
% X(x)X(x) Terms (Evaluated part)
Krotx0XmXi=zeros(nmax);
Krotx0XmXi(2,2)=1;
% Y(y)Y(y) Terms (Integral part)
Krotx0YnYj=X00;
for m=1:nmax
    for i=1:nmax
        u=i+(m-1)*nmax;
        for n=1:nmax
            for j=1:nmax
                v=j+(n-1)*nmax;
                Krotx0(u,v)=...
                Kr1*Krotx0XmXi(m,n)* ...
                Krotx0YnYj(i,j)+ ...
                Kr2*Krotx0YnYj(m,n)* ...
                Krotx0XmXi(i,j);
            end
        end
    end
end
%%%%%%%%%%%%%%%%%%%%%%%%%%%%%%%%%%%%%%%%%%%%%%%%%%%%%%%%
%              EDGE3 & EDGE4 (X=1 & Y=1)
%%%%%%%%%%%%%%%%%%%%%%%%%%%%%%%%%%%%%%%%%%%%%%%%%%%%%%%%
% X(x)X(x) Terms (Evaluated part)
Krotx1XmXi=zeros(nmax);
Krotx1XmXi(2,2)=1;
Krotx1XmXi(2,3)=2;
Krotx1XmXi(3,2)=2;
Krotx1XmXi(3,3)=4;
% Y(y)Y(y) Terms (Integral part)
Krotx1YnYj=X00;
for m=1:nmax
    for i=1:nmax
```

```
            u=i+(m-1)*nmax;
            for n=1:nmax
                for j=1:nmax
                    v=j+(n-1)*nmax;
                    Krotx1(u,v)= ...
                    Kr3*Krotx1XmXi(m,n)* ...
                    Krotx1YnYj(i,j)+...
                    Kr4*Krotx1YnYj(m,n)* ...
                    Krotx1XmXi(i,j);
                end
            end
        end
end
KPPUUandVV=(Ktransx0+Ktransx1);
KPPWW=(Krotx0+Krotx1+Ktransx0+Ktransx1);
Penalty_Matrix=[KPPUUandVV ZEROS ZEROS; ...
    ZEROS KPPUUandVV ZEROS; ZEROS ZEROS KPPWW];
%%%%%%%%%%%%%%%%%%%%%%%%%%%%%%%%%%%%%%%%%%%%%%%%%%%%%%%%%%%%%%%%%%%%%%%%%%%%%%
%STIFFNESS MATRIX AS FUNCTION OF GEOMETRIC PORPERTIES
%%%%%%%%%%%%%%%%%%%%%%%%%%%%%%%%%%%%%%%%%%%%%%%%%%%%%%%%%%%%%%%%%%%%%%%%%%%%%%
for m=1:nmax
    for i=1:nmax
        u=i+(m-1)*nmax;
        for n=1:nmax
            for j=1:nmax
                v=j+(n-1)*nmax;
                UU(u,v) = ...
                (b/h)^2 * X11(m,n) * Y00(i,j) ...
                + 1/2 * (1-poisson) * (a/h)^2 * ...
                X00(m,n) * Y11(i,j);
                UV(u,v) =  poisson * (a/h) *(b/h)* ...
                X10(m,n) * Y01(i,j) + ...
                1/2 * (1-poisson) * (a/h) * ...
                (b/h) * X01(m,n) * Y10(i,j);
                UW(u,v) = (a/h) * (b/h) * ...
                ((b/Rx)+((poisson*b)/(Ry)))* ...
                X10(m,n) * Y00(i,j);
                VV(u,v) = ...
                (a/h)^2 * X00(m,n) * Y11(i,j) + ...
                1/2 * (1-poisson) * (b/h)^2 * ...
                X11(m,n) * Y00(i,j);
                VW(u,v) = ...
                (a/h)*(b/h)*(((poisson*a)/(Rx))+ ...
                (a/(Ry)))*X00(m,n)*Y10(i,j);
                WW(u,v)= (a/h)^2 * ...
```

```
                    ((b/(Rx))^2 + 2*poisson*(b/(Rx))* ...
                    (b/(Ry)) + (b/(Ry))^2 ) * ...
                    X00(m,n) * Y00(i,j) +...
                    1/12*(b/a)^2*X22(m,n)*Y00(i,j) + ...
                    1/12*(a/b)^2*X00(m,n)*Y22(i,j) +...
                    1/12*poisson*X02(m,n)*Y20(i,j) +...
                    1/12*poisson*X20(m,n)*Y02(i,j) +...
                    2/12*(1-poisson)*X11(m,n)*Y11(i,j);
               end
          end
     end
end
VU=UV';
WU=UW';
WV=VW';
K = 12*[UU UV UW; VU VV VW; WU WV WW];
%%%%%%%%%%%%%%%%%%%%%%%%%%%%%%%%%%%%%%%%%%%%%%%%%%%%%%%%%%%%%%
KTOT=K+Penalty_Matrix*PenaltyParameter;
[X,Eigenvalues]= eig(KTOT,M);
Eigenvalues=sort(diag(Eigenvalues));
Non_Dim_Freq_Parameter=real(Eigenvalues.^0.5)


% MODES FOR A CONICAL SHELL

close all
% MODES FOR SPHERICAL SHELLS ONLY
Mode_No=9;
%for MODE_NUMBER=Mode_No:Mode_No;
step_x=1/49; step_y=1/49;
% SET_X
x_points=0:step_x:1;
y_points=0:step_y:1;
steps_X=size(x_points,2);
steps_Y=size(y_points,2);
for MODE_NUMBER=1:Mode_No
EIGENVECTOR=real(X(:,MODE_NUMBER))';
for i=1:steps_X
    x=x_points(i);
    for j=1:steps_Y
         y=y_points(j);
% SET_X and SET_Y
        SET_X(1)=1;
        SET_X(2)=x;
        SET_X(3)=x^2;
```

```
        SET_Y(1)=1;
        SET_Y(2)=y;
        SET_Y(3)=y^2;
        for k=1:nmax-3
            SET_X(k+3)=cos(k*pi*x);
            SET_Y(k+3)=cos(k*pi*y);
        end
        counter=0;
        for n=1:nmax
            for m=1:nmax
                counter=counter+1;
                EVALUATED_FUNCTIONS(counter)= ...
                SET_X(m)*SET_Y(n);
            end
        end
        EIGENVECTOR_TEMP= ...
        reshape(EIGENVECTOR, nmax^2, 3);
        EIGENVECTORX=  EIGENVECTOR_TEMP(:,1);
        EIGENVECTORY=  EIGENVECTOR_TEMP(:,2);
        EIGENVECTORZ=  EIGENVECTOR_TEMP(:,3);
        EVALUATED_MODE_MODE_X(i,j)= ...
        dot(EVALUATED_FUNCTIONS,EIGENVECTORX);
        EVALUATED_MODE_MODE_Y(i,j)= ...
        dot(EVALUATED_FUNCTIONS,EIGENVECTORY);
        EVALUATED_MODE_MODE_Z(i,j)= ...
        dot(EVALUATED_FUNCTIONS,EIGENVECTORZ);
    end
end
%%%%%%%%%%%%%%%%%%%%%%%%%%%%%%%%%%%%%%%%%%%%%
figure1 = figure('Color',[1 1 1]);
set(gca,'Units','normalized','Position',[0 0 1 1]);
set(gcf,'Units','pixels','Position',[50 50 600 600]);
hold on;
axes1 = axes('Parent',figure1);
hold(axes1,'all');
view(axes1,[75 45]);
for i=1:size(x_points,2)
    for j=1:size(y_points,2)
        z_points(i,j)= ...
        sqrt(Rx^2-(x_points(i)-0.5)^2- ...
       (y_points(j)-0.5)^2);
    end
end
X_PLOT=[];
Y_PLOT=[];
```

```
for i=1:size(y_points,2)
    X_PLOT=[X_PLOT; x_points];
    Y_PLOT=[Y_PLOT; y_points];
end
Z_PLOT=z_points+EVALUATED_MODE_MODE_Z;
surf(x_points,x_points,Z_PLOT)
end
```

Index

A, B, C

D, E, F

G, I, K, L

M, N, O

P, R, S

T, U, W

Other titles from

iSTE

in

Mechanical Engineering and Solid Mechanics

2014

ATANACKOVIC M. Teodor, PILIPOVIC Stevan, STANKOVIC Bogoljub, ZORICA Dusan
Fractional Calculus with Applications in Mechanics: Vibrations and Diffusion Processes

ATANACKOVIC M. Teodor, PILIPOVIC Stevan, STANKOVIC Bogoljub, ZORICA Dusan
Fractional Calculus with Applications in Mechanics: Wave Propagation, Impact and VariationalPrinciples

CIBLAC Thierry, MOREL Jean-Claude
Sustainable Masonry: Stability and Behavior of Structures

LALANNE Christian
Mechanical Vibration and Shock Analysis – 5-volume series – 3rd edition
Sinusoidal Vibration – volume 1
Mechanical Shock – volume 2
Random Vibration – volume 3
Fatigue Damage – volume 4
Specification Development – volume 5

LEMAIRE Maurice
Uncertainty and Mechanics

2013

ADHIKARI Sondipon
Structural Dynamic Analysis with Generalized Damping Models: Identification

ADHIKARI Sondipon
Structural Dynamic Analysis with Generalized Damping Models: Analysis

BAILLY Patrice
Materials and Structures under Shock and Impact

BASTIEN Jérôme, BERNARDIN Frédéric, LAMARQUE Claude-Henri
Non-smooth Deterministic or Stochastic Discrete Dynamical Systems: Applications to Models with Friction or Impact

EL HAMI Abdelkhalak, BOUCHAIB Radi
Uncertainty and Optimization in Structural Mechanics

KIRILLOV Oleg N., PELINOVSKY Dmitry E.
Nonlinear Physical Systems: Spectral Analysis, Stability and Bifurcations

LUONGO Angelo, ZULLI Daniele
Mathematical Models of Beams and Cables

SALENÇON Jean
Yield Design

2012

DAVIM J. Paulo
Mechanical Engineering Education

DUPEUX Michel, BRACCINI Muriel
Mechanics of Solid Interfaces

ELISHAKOFF Isaac *et al.*
Carbon Nanotubes and Nanosensors: Vibration, Buckling and Ballistic Impact

GRÉDIAC Michel, HILD François
Full-Field Measurements and Identification in Solid Mechanics

GROUS Ammar
Fracture Mechanics – 3-volume series
Analysis of Reliability and Quality Control – volume 1
Applied Reliability – volume 2
Applied Quality Control – volume 3

RECHO Naman
Fracture Mechanics and Crack Growth

2011

KRYSINSKI Tomasz, MALBURET François
Mechanical Instability

SOUSTELLE Michel
An Introduction to Chemical Kinetics

2010

BREITKOPF Piotr, FILOMENO COELHO Rajan
Multidisciplinary Design Optimization in Computational Mechanics

DAVIM J. Paulo
Biotribolgy

PAULTRE Patrick
Dynamics of Structures

SOUSTELLE Michel
Handbook of Heterogenous Kinetics

2009

BERLIOZ Alain, TROMPETTE Philippe
Solid Mechanics using the Finite Element Method

LEMAIRE Maurice
Structural Reliability

2007

GIRARD Alain, ROY Nicolas
Structural Dynamics in Industry

GUINEBRETIÈRE René
X-ray Diffraction by Polycrystalline Materials

KRYSINSKI Tomasz, MALBURET François
Mechanical Vibrations

SIH George C. *et al.*
Particle and Continuum Aspects of Mesomechanics

Printed and bound by CPI Group (UK) Ltd, Croydon, CR0 4YY

13/09/2026

14961079-0003